INTRODUCTION TO DIGITAL SIGNAL PROCESSING

Edited by
Galang, Maureen R.

Delve Publishing

Introduction to Digital Signal Processing

Edited by Galang, Maureen R.

Edition 2018

Delve Publishing

Published by
Delve Publishing LLC
708 3rd Avenue, 6th Floor
New York
NY 10801
United States of America
www.delvepublishing.com

ISBN: 978-1-68095-904-8 (Hardback)

Printed in United States of America

Library of Congress Control Number: 2016920463

About the Editor

Galang, Maureen R.

Maureen Reyes Galang received her Bachelor's degree in Electrical Engineering from Technological University of the Philippines, Manila in 2010. During her first career, Engr. Galang has been professionally involved in producing high quality 3D models, i.e. electrical equipment and devices for automotive and other machineries. Later of her profession focuses on building construction, i.e. residential, commercial and industrial. She became a highly skilled Electrical Engr. in Quantity Surveying and Cost Analysis bringing cost wise bids for clients.

Table of Contents

5 THEORY OF SAMPLING

6 DATA COMPRESSION

7 KALMAN FILTERING

8 DIGITAL SIGNAL PROCESSOR

9 FINITE IMPULSE RESPONSE

10 CONVOLUTION APPLICATIONS TO DSP

List of Abbreviations

A/D	Analog-to-digital
AAF	Anti-aliasing filter
ADC	Analog-to-digital converter
ASICs	Application-specific integrated circuit
ASP	Analog signal processing
AWGN	Additive white Gaussian noise
BLSE	Bayes least squared error
BLUE	Best linear unbiased estimator
CD	Compact disc
CISC	Complex instruction set computing
CMYK	Cyan, magenta, yellow, and black
CRLB	Cramér–Rao lower bound
DAC	Digital-to-analog converter
DCT	Discrete cosine transform
DFT	Discrete Fourier transform
DSP	Digital signal processor
DWT	Discrete wavelet transform
ECL	Emitter-coupled logic
EPIC	Explicitly parallel instruction computing
FIR	Finite impulse response
FM	Frequency modulation
FPGAs	Field-programmable gate arrays
GPUs	Graphics processing units

HMM	hidden Markov model
IIR	Infinite impulse response
ILP	Instruction level parallelism
LTI	Linear time-invariant LTI
LO	Local oscillator LO
LPC	Linear predictive coding
LQE	Linear quadratic estimation
LQG `	Linear-quadratic-Gaussian control problem
MCMC	Markov chain Monte Carlo
MAP	Maximum a posteriori
MSI	Medium-scale integration
MMSE	Minimum mean squared error
MVUE	Minimum variance unbiased estimator
MBF	modified Bryson–Frazier
MDCT	Modified discrete cosine transform
MVI	Motion Video Instructions
MAC	Multiply–accumulate
RAM	Random-access memory
SDT	Signal detection theory
SIMD	Single instruction, multiple data
SRAM	Static random-access memory
SHARC	Super Harvard Architecture Single-Chip Computer
TI	Texas Instruments
TDP	Typical thermal design power
VLSI	Very large-scale integration
VLIW	Very long instruction word
WKS	Whittaker–Kotel'nikov–Shannon

Preface

The world of science and engineering is filled with signals: images from remote space probes, voltages generated by the heart and brain, radar and sonar echoes, seismic vibrations, and countless other applications. Digital Signal Processing (DSP) is the science of using computers to understand these types of data. It refers to various techniques for improving the accuracy and reliability of digital communications. The theory behind DSP is quite complex. DSP is primarily used to detect errors, and to filter and compress analog signals in transit. Basically, DSP works by clarifying, or standardizing, the levels or states of a digital signal. ADSP circuit is able to differentiate between human-made signals, which are orderly, and noise, which is inherently chaotic. Digital Signal Processors (DSP) take real-world signals like voice, audio, video, temperature, pressure, or position that have been digitized and then mathematically manipulate them. Signals need to be processed so that the information that they contain can be displayed, analyzed, or converted to another type of signal that may be of use. In the real-world, analog products detect signals such as sound, light, temperature or pressure and manipulate them. Converters such as an Analog-to-Digital converter then take the real-world signal and turn it into the digital format of 1's and 0's. From here, the DSP takes over by capturing the digitized information and processing it. It then feeds the digitized information back for use in the real world. It does this in one of two ways, either digitally or in an analog format by going through a Digital-to-Analog converter. All of this occurs at very high speeds. DSP is one of the most powerful technologies that will shape science and engineering in the twenty-first century. This text places the emphasis on understanding the principles and applications of DSP.

Content Coverage

Chapters One and *Two* present an overview of DSP and theory of DSP, respectively. Signal detection theory provides a precise language and graphic notation for analyzing decision making in the presence of uncertainty. The general approach of signal detection theory has direct application for us in terms of sensory experiments.

Chapters Three and *Four* provide coverage of sub-fields of digital signal processing and advanced Z-transform, separately. In mathematics and signal processing, the advanced Z-transform is an extension of the Z-transform, to incorporate ideal delays that are not multiples of the sampling time.

Chapters Five and *Six* focus on theory of sampling and data compression, correspondingly. In signal processing, sampling is the reduction of a continuous-time signal to a discrete-time signal and compression is useful because it reduces resources required to store and transmit data.

Chapters Seven and *Eight* present about Kalman filtering approaches, and digital signal processor; a specialized microprocessor, with its architecture optimized for the operational needs of digital signal processing, separately.

Chapters Nine and *Ten* are FIR filters and applications of convolution to DSP, distinctly. In signal processing, a finite impulse response (FIR) filter is a filter whose impulse response is of finite duration, because it settles to zero in finite time.

CHAPTER 1

Digital Signal Processing

Chapter Contents

- **Introduction**
- **Signal Sampling**
- **Domains**
- **Applications**
- **Implementation**
- **Frequency Domain**

Digital signal processing (DSP) refers to various techniques for improving the accuracy and reliability of digital communications. The theory behind DSP is quite complex. Basically, DSP works by clarifying, or standardizing, the levels or states of a digital signal. ADSP circuit is able to differentiate between human-made signals,which are orderly, and noise, which is inherently chaotic.

INTRODUCTION

Digital signal processing (**DSP**) is the use of digital processing, such as by computers, to perform a wide variety of signal processing operations. The signals processed in this manner are a sequence of numbers that represent samples of a continuous variable in a domain such as time, space, or frequency.

Digital signal processing and analog signal processing are subfields of signal processing. DSP applications include audio and speech signal processing, sonar, radar and other sensor array processing, spectral estimation, statistical signal processing, digital image processing, signal processing for telecommunications, control of systems, biomedical engineering, seismic data processing, among others.

Digital signal processing can involve linear or nonlinear operations. Nonlinear signal processing is closely related to nonlinear system identification and can be implemented in the time, frequency, and spatio-temporal domains.

The application of digital computation to signal processing allows for many advantages over analog processing in many applications, such as error detection and correction in transmission as well as data compression. DSP is applicable to both streaming data and static (stored) data.

SIGNAL SAMPLING

The increasing use of computers has resulted in the increased use of, and need for, digital signal processing. To digitally analyze and manipulate an analog signal, it must be digitized with an analog-to-digital converter. Sampling is usually carried out in two stages, discretization and quantization. Discretization means that the signal is divided into equal intervals of time, and each interval is represented by a single measurement of amplitude. Quantization means each amplitude measurement is approximated by a value from a finite set. Rounding real numbers to integers is an example.

The Nyquist–Shannon sampling theorem states that a signal can be exactly reconstructed from its samples if the sampling frequency is greater than twice the highest frequency of the signal. In practice, the sampling frequency is often significantly higher than twice that required by the signal's limited bandwidth.

Theoretical DSP analyses and derivations are typically performed on discrete-time signal models with no amplitude inaccuracies (quantization error), created by the abstract process of sampling. Numerical methods require a quantized signal, such as those produced by an analog-to-digital converter (ADC). The processed result might be a frequency spectrum or a set of statistics. But often it is another quantized signal that is converted back to analog form by a digital-to-analog converter (DAC).

DOMAINS

In DSP, engineers usually study digital signals in one of the following domains: time domain (one-dimensional signals), spatial domain (multidimensional signals), frequency domain, and wavelet domains. They choose the domain in which to process a signal by making an informed assumption (or by trying different possibilities) as to which domain best represents the essential characteristics of the signal. A sequence of samples from a measuring device produces a temporal or spatial domain representation, whereas a discrete Fourier transform produces the frequency domain information, that is, the frequency spectrum.

Time and Space Domains

The most common processing approach in the time or space domain is enhancement of the input signal through a method called filtering. Digital filtering generally consists of some linear transformation of a number of surrounding samples around the current sample of the input or output signal. There are various ways to characterize filters; for example:

- A linear filter is a linear transformation of input samples; other filters are "non-linear". Linear filters satisfy the superposition condition, i.e. if an input is a weighted linear combination of different signals, the output is a similarly weighted linear combination of the corresponding output signals.
- A causal filter uses only previous samples of the input or output signals; while a non-causal filter uses future input samples. A non-causal filter can usually be changed into a causal filter by adding a delay to it.
- A time-invariant filter has constant properties over time; other filters such as adaptive filters change in time.
- A stable filter produces an output that converges to a constant value with time, or remains bounded within a finite interval. An unstable filter can produce an output that grows without bounds, with bounded or even zero input.
- A finite impulse response (FIR) filter uses only the input signals, while an infinite impulse response filter (IIR) uses both the input signal and previous samples of the output signal. FIR filters are always stable, while IIR filters may be unstable.

A filter can be represented by a block diagram, which can then be used to derive a sample processing algorithm to implement the filter with hardware instructions. A filter may also be described as a difference equation, a collection of zeroes and poles or, if it is an FIR filter, an impulse response or step response.

The output of a linear digital filter to any given input may be calculated by convolving the input signal with the impulse response.

Frequency domain

Signals are converted from time or space domain to the frequency domain usually through the Fourier transform. The Fourier transform converts the signal information to a magnitude and phase component of each frequency. Often the Fourier transform is converted to the power spectrum, which is the magnitude of each frequency component squared.

The most common purpose for analysis of signals in the frequency domain is analysis of signal properties. The engineer can study the spectrum to determine which frequencies are present in the input signal and which are missing.

In addition to frequency information, phase information is often needed. This can be obtained from the Fourier transform. With some applications, how the phase varies with frequency can be a significant consideration.

Filtering, particularly in non-real-time work can also be achieved by converting to the frequency domain, applying the filter and then converting back to the time domain. This is a fast, O(n log n) operation, and can give essentially any filter shape including excellent approximations to brickwall filters.

There are some commonly used frequency domain transformations. For example, the cepstrum converts a signal to the frequency domain through Fourier transform, takes the logarithm, then applies another Fourier transform. This emphasizes the harmonic structure of the original spectrum.

Frequency domain analysis is also called spectrum- or spectral analysis.

Z-plane Analysis

Digital filters come in both IIR and FIR types. FIR filters have many advantages, but are computationally more demanding. Whereas FIR filters are always stable, IIR filters have feedback loops that may resonate when stimulated with certain input signals. The Z-transform provides a tool for analyzing potential stability issues of digital IIR filters. It is analogous to the Laplace transform, which is used to design analog IIR filters.

Wavelet

An example of the 2D discrete wavelet transform that is used in JPEG2000. The original image is high-pass filtered, yielding the three large images, each describing local changes in brightness (details) in the original image. It is then low-pass filtered and downscaled, yielding an approximation image; this image is high-pass filtered to produce the three smaller detail images, and low-pass filtered to produce the final approximation image in the upper-left.

In numerical analysis and functional analysis, a discrete wavelet transform (DWT) is any wavelet transform for which the wavelets are discretely sampled. As with other wavelet transforms, a key advantage it has over Fourier transforms is temporal resolution: it captures both frequency and location information.[a]

APPLICATIONS

The main applications of DSP are audio signal processing, audio compression, digital image processing, video compression, speech processing, speech recognition, digital communications, digital synthesizers, radar, sonar, financial signal processing, seismology and biomedicine. Specific examples are speech compression and transmission in digital mobile phones, room correction of sound in hi-fi and sound reinforcement applications, weather forecasting, economic forecasting, seismic data processing, analysis and control of industrial processes, medical imaging such as CAT scans and MRI, MP3 compression, computer graphics, image manipulation, hi-fi loudspeaker crossovers and equalization, and audio effects for use with electric guitar amplifiers.

IMPLEMENTATION

DSP algorithms have long been run on general-purpose computers and digital signal processors. DSP algorithms are also implemented on purpose-built hardware such as application-specific integrated circuit (ASICs). Additional technologies for digital signal processing include more powerful general purpose microprocessors, field-programmable gate arrays (FPGAs), digital signal controllers (mostly for industrial applications such as motor control), and stream processors.

Depending on the requirements of the application, digital signal processing tasks can be implemented on general purpose computers.

Often when the processing requirement is not real-time, processing is economically done with an existing general-purpose computer and the signal data (either input or output) exists in data files. This is essentially no different from any other data processing, except DSP mathematical techniques (such as the FFT) are used, and the sampled data is usually assumed to be uniformly sampled in time or space. For example: processing digital photographs with software such as Photoshop.

However, when the application requirement is real-time, DSP is often implemented using specialized microprocessors such as the DSP56000, the TMS320, or the SHARC. These often process data using fixed-point arithmetic, though some more powerful versions use floating point. For faster applications FPGAs might be used. Beginning in 2007, multicore implementations of DSPs have started to emerge from companies including Freescale and Stream Processors, Inc. For faster applications with vast usage, ASICs might be designed specifically. For slow applications, a traditional slower processor such as a microcontroller may be adequate. Also a growing number of DSP applications are now being implemented on embedded systems using powerful PCs with multi-core processors

FREQUENCY DOMAIN

The Fourier transform relates the function's time domain, shown in red, to the function's frequency domain, shown in blue. The component frequencies, spread across the frequency spectrum, are represented as peaks in the frequency domain.

In electronics, control systems engineering, and statistics, the frequency domain refers to the analysis of mathematical functions or signals with respect to frequency, rather than time. Put simply, a time-domain graph shows how a signal changes over time, whereas a frequency-domain graph shows how much of the signal lies within each given frequency band over a range of frequencies. A frequency-domain representation can also include information on the phase shift that must be applied to each sinusoid in order to be able to recombine the frequency components to recover the original time signal.

A given function or signal can be converted between the time and frequency domains with a pair of mathematical operators called a transform. An example is the Fourier transform, which converts the time function into a sum of sine waves of different frequencies, each of which represents a frequency component. The 'spectrum' of frequency components is the frequency domain representation of the signal. The inverse Fourier transform converts the frequency domain function back to a time function. A spectrum analyzer is the tool commonly used to visualize real-world signals in the frequency domain.

Some specialized signal processing techniques use transforms that result in a joint time-frequency domain, with the instantaneous frequency being a key link between the time domain and the frequency domain.

Magnitude and Phase

In using the Laplace, Z-, or Fourier transforms, a signal is described by a complex function of frequency: the component of the signal at any given frequency is given by a complex number. The magnitude of the number is the amplitude of that component, and the angle is the relative phase of the wave. For example, using the Fourier transform a sound wave, such as human speech, can be broken down into its component tones of different frequencies, each represented by a sine wave of a different amplitude and phase. The response of a system, as a function of frequency, can also be described by a complex function. In many applications, phase information is not important. By discarding the phase information it is possible to simplify the information in a frequency domain representation to generate a frequency spectrum or spectral density. A spectrum analyzer is a device that displays the spectrum, while the time domain frequency can be seen on an oscilloscope.

The power spectral density is a frequency-domain description that can be applied to a large class of signals that are neither periodic nor square-integrable; to have a power spectral density, a signal needs only to be the output of a wide-sense stationary random process.

Different Frequency Domains

Although "*the*" frequency domain is spoken of in the singular, there are a number of different mathematical transforms which are used to analyze time domain functions and are referred to as "frequency domain" methods. These are the most common transforms, and the fields in which they are used:

- Fourier series – repetitive signals, oscillating systems
- Fourier transform – nonrepetitive signals, transients
- Laplace transform – electronic circuits and control systems
- Z transform – discrete signals, digital signal processing
- Wavelet transform - image analysis, data compression

More generally, one can speak of the transform domain with respect to any transform. The above transforms can be interpreted as capturing some form of frequency, and hence the transform domain is referred to as a frequency domain.

Discrete Frequency Domain

The Fourier transform of a periodic signal only has energy at a base frequency and its harmonics. Another way of saying this is that a periodic signal can be analyzed using a discrete frequency domain. Dually, a discrete-time signal gives rise to a periodic frequency spectrum. Combining these two, if we start with a time signal which is both discrete and periodic, we get a frequency spectrum which is both periodic and discrete. This is the usual context for a discrete Fourier transform.

History of Term

The use of the terms "frequency domain" and "time domain" arose in communication engineering in the 1950s and early 1960s, with "frequency domain" appearing in 1953. See time domain: origin of term for details.

REFERENCES

1. Billings, Stephen A. (Sep 2013). Nonlinear System Identification: NARMAX Methods in the Time, Frequency, and Spatio-Temporal Domains. UK: Wiley. ISBN 978-1-119-94359-4.
2. Broesch, James D.; Stranneby, Dag; Walker, William (2008-10-20). Digital Signal Processing: Instant access (1 ed.). Butterworth-Heinemann - Newnes. p. 3. ISBN 9780750689762.
3. Stranneby, Dag; Walker, William (2004). Digital Signal Processing and Applications (2nd ed.). Elsevier. ISBN 0-7506-6344-8.
4. JpFix (2006). FPGA-Based Image Processing Accelerator. Retrieved 2008-05-10.

CHAPTER 2

Theory of Digital Signal Processing

Chapter Contents

The starting point for signal detection theory is that nearly all reasoning and decision making takes place in the presence of some uncertainty. Signal detection theory provides a precise language and graphic notation for analyzing decision making in the presence of uncertainty. The general approach of signal detection theory has direct application for us in terms of sensory experiments. But it also offers a way to analyze many different kinds of decision problems.

DETECTION THEORY

Detection theory or signal detection theory is a means to quantify the ability to discern between information-bearing patterns (called stimulus in living organisms, signal in machines) and random patterns that distract from the information (called noise, consisting of background stimuli and random activity of the detection machine and of the nervous system of the operator). In the field of electronics, the separation of such patterns from a disguising background is referred to as signal recovery.

According to the theory, there are a number of determiners of how a detecting system will detect a signal, and where its threshold levels will be. The theory can explain how changing the threshold will affect the ability to discern, often exposing how adapted the system is to the task, purpose or goal at which it is aimed.

When the detecting system is a human being, experience, expectations, physiological state (e.g., fatigue) and other factors can affect the threshold applied. For instance, a sentry in wartime might be likely to detect fainter stimuli than the same sentry in peacetime due to a lower criterion, however they might also be more likely to treat innocuous stimuli as a threat.

Much of the early work in detection theory was done by radar researchers. By 1954, the theory was fully developed on the theoretical side as described by Peterson, Birdsall and Fox and the foundation for the psychological theory was made by Wilson P. Tanner, David M. Green, and John A. Swets, also in 1954.Detection theory was used in 1966 by John A. Swets and David M. Green for psychophysics. Green and Swets criticized the traditional methods of psychophysics for their inability to discriminate between the real sensitivity of subjects and their (potential) response biases.

Detection theory has applications in many fields such as diagnostics of any kind, quality control, telecommunications, and psychology. The concept is similar to the signal to noise ratio used in the sciences and confusion matrices used in artificial intelligence. It is also usable in alarm management, where it is important to separate important events from background noise.

PSYCHOLOGY

Signal detection theory (SDT) is used when psychologists want to measure the way we make decisions under conditions of uncertainty, such as how we would perceive distances in foggy conditions. SDT assumes that the decision maker is not a passive receiver of information, but an active decision-maker who makes difficult perceptual judgments under conditions of uncertainty. In foggy circumstances, we are forced to decide how far away from us an object is, based solely upon visual stimulus which is impaired by the fog. Since the brightness of the object, such as a traffic light, is used by the brain to discriminate the distance of an object, and the fog reduces the brightness of objects, we perceive the object to be much farther away than it actually is (see also decision theory).

To apply signal detection theory to a data set where stimuli were either present or absent, and the observer categorized each trial as having the stimulus present or absent, the trials are sorted into one of four categories:

	Respond "Absent"	**Respond "Present"**
Stimulus Present	Miss	Hit
Stimulus Absent	Correct Rejection	False Alarm

Based on the proportions of these types of trials, numerical estimates of sensitivity can be obtained with statistics like the sensitivity index *d'* and A', and response bias can be estimated with statistics like c and β.

Signal detection theory can also be applied to memory experiments, where items are presented on a study list for later testing. A test list is created by combining these 'old' items with novel, 'new' items that did not appear on the study list. On each test trial the subject will respond 'yes, this was on the study list' or 'no, this was not on the study list'. Items presented on the study list are called Targets, and new items are called Distractors. Saying 'Yes' to a target constitutes a Hit, while saying 'Yes' to a distractor constitutes a False Alarm.

	Respond "No"	Respond "Yes"
Target	Miss	Hit
Distractor	Correct Rejection	False Alarm

APPLICATIONS

Signal Detection Theory has wide application, both in humans and other animals. Topics include memory, stimulus characterists of schedules of reinforcement, etc.

Sensitivity or discriminability

Conceptually, sensitivity refers to how hard or easy it is to detect that a target stimulus is present from background events. For example, in a recognition memory paradigm, having longer to study to-be-remembered words makes it easier to recognize previously seen or heard words. In contrast, having to remember 30 words rather than 5 makes the discrimination harder. One of the most commonly used statistics for computing sensitivity is the so-called sensitivity index or d'. There are also non-parametric measures, such as the area under the ROC-curve.

Bias

Bias is the extent to which one response is more probable than another. That is, a receiver may be more likely to respond that a stimulus is present or more likely to respond that a stimulus is not present. Bias is independent of sensitivity. For example, if there is a penalty for either false alarms or misses, this may influence bias. If the stimulus is a bomber, then a miss (failing to detect the plane) may increase deaths, so a liberal bias is likely. In contrast, crying wolf (a false alarm) too often may make people less likely to respond, grounds for a conservative bias.

MATHEMATICS

P(H1|y) > P(H2|y) / MAP Testing

In the case of making a decision between two hypotheses, *H1*, absent, and *H2*, present, in the event of a particular observation, *y*, a classical approach is to choose *H1* when $p(H1|y) > p(H2|y)$ and *H2* in the reverse case. In the event that the two *a posteriori* probabilities are equal, one might choose to default to a single choice (either always choose *H1* or always choose *H2*), or might randomly select either *H1* or *H2*. The *a priori* probabilities of *H1* and *H2* can guide this choice, e.g. by always choosing the hypothesis with the higher *a priori* probability.

When taking this approach, usually what one knows are the conditional probabilities, $p(H1) = \pi_1$ and $p(H2) = \pi_2$ In this case,

$$p(H1|y) = \frac{p(y|H1) \cdot \pi_1}{p(y)},$$

$$p(H2|y) = \frac{p(y|H2) \cdot \pi_2}{p(y)}$$

where p(y) is the total probability of event y,

$p(y|H1) \cdot \pi_1 + p(y|H2) \cdot \pi_2$.

H2 is chosen in case

$$\frac{p(y|H2) \cdot \pi_2}{p(y|H1) \cdot \pi_1 + p(y|H2) \cdot \pi_2} \geq \frac{p(y|H1) \cdot \pi_1}{p(y|H1) \cdot \pi_1 + p(y|H2) \cdot \pi_2}$$

$$\Rightarrow \frac{p(y|H2)}{p(y|H1)} \geq \frac{\pi_1}{\pi_2}$$

and H1 otherwise.

Often, the ratio $\frac{\pi_1}{\pi_2}$ is called τ_{MAP} and $\frac{p(y|H2)}{p(y|H1)}$ is called $L(y)$, the likelihood ratio.
Using this terminology, H2 is chosen in case. This is called MAP testing, where MAP stands for "maximum a posteriori").

Taking this approach minimizes the expected number of errors one will make.

Bayes Criterion

In some cases, it is far more important to respond appropriately to H1 than it is to respond appropriately to H2. For example, if an alarm goes off, indicating H1 (an incoming bomber is carrying a nuclear weapon), it is much more important to shoot down the bomber if H1 = TRUE, than it is to send a fighter squadron to inspect a false alarm (i.e., H1 = FALSE, H2 = TRUE) (assuming a large supply of fighter squadrons). The Bayes criterion is an approach suitable for such cases.

Here a utility is associated with each of four situations:

- U_{11}: One responds with behavior appropriate to H1 and H1 is true: fighters destroy bomber, incurring fuel, maintenance, and weapons costs, take risk of some being shot down;
- U_{12}: One responds with behavior appropriate to H1 and H2 is true: fighters sent out, incurring fuel and maintenance costs, bomber location remains unknown;
- U_{21}: One responds with behavior appropriate to H2 and H1 is true: city destroyed;
- U_{22}: One responds with behavior appropriate to H2 and H2 is true: fighters stay home, bomber location remains unknown;

As is shown below, what is important are the differences, $U_{11} - U_{21}$ and $U_{22} - U_{12}$.

Similarly, there are four probabilities, P_{11}, P_{12}, etc., for each of the cases (which are dependent on one's decision strategy).

The Bayes criterion approach is to maximize the expected utility:

$$U = P_{11} \cdot U_{11} + P_{21} \cdot U_{21} + P_{12} \cdot U_{12} + P_{22} \cdot U_{22}$$

$$U = P_{11} \cdot U_{11} + (1 - P_{11}) \cdot U_{21} + P_{12} \cdot U_{12} + (1 - P_{12}) \cdot U_{22}$$

$$U = U_{21} + U_{22} + P_{11} \cdot (U_{11} - U_{21}) - P_{12} \cdot (U_{22} - U_{12})$$

Effectively, one may maximize the sum,

$$U' = P_{11} \cdot (U_{11} - U_{21}) - P_{12} \cdot (U_{22} - U_{12}),$$

and make the following substitutions:

$$P_{11} = \pi_1 \cdot \int_{R_1} p(y|H1)\, dy$$

$$P_{12} = \pi_2 \cdot \int_{R_1} p(y|H2)\, dy$$

where π_1 and π_2 are the a priori probabilities, $P(H1)$ and $P(H2)$, and R_1, is the region of observation events, y, that are responded to as though H1 is true.

$$\Rightarrow U' = \int_{R_1} \{\pi_1 \cdot (U_{11} - U_{21}) \cdot p(y|H1) - \pi_2 \cdot (U_{22} - U_{12}) \cdot p(y|H2)\}\, dy$$

U' and thus U are maximized by extending R_1 over the region where

$$\pi_2 \cdot (U_{22} - U_{12}) \cdot p(y|H2) \geq \pi_1 \cdot (U_{11} - U_{21}) \cdot p(y|H1)$$

$$\Rightarrow L(y) \equiv \frac{p(y|H2)}{p(y|H1)} \geq \frac{\pi_1 \cdot (U_{11} - U_{21})}{\pi_2 \cdot (U_{22} - U_{12})} \equiv \tau_B$$

and H1 otherwise, where L(y) is the so-defined likelihood ratio.

Discrete-Time Signal

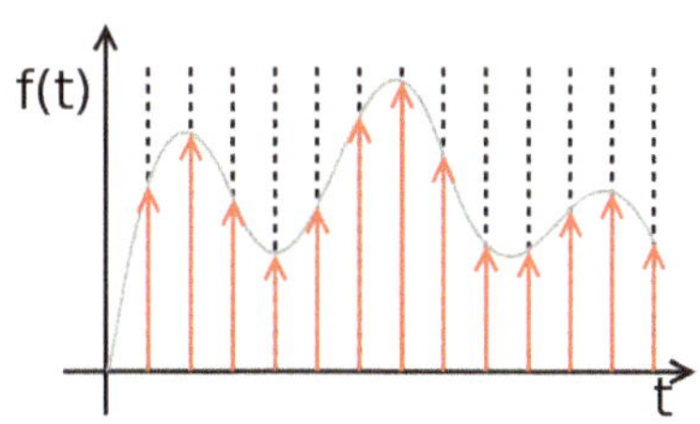

Discrete sampled signal

"Discrete signal" redirects here. It is not to be confused with Digital signal (electronics).

A discrete signal or discrete-time signal is a time series consisting of a sequence of quantities. In other words, it is a time series that is a function over a domain of integers.

Unlike a continuous-time signal, a discrete-time signal is not a function of a continuous argument; however, it may have been obtained by sampling from a continuous-time signal, and then each value in the sequence is called a sample. When a discrete-time signal obtained by sampling a sequence corresponds to uniformly spaced times, it has an associated sampling rate; the sampling rate is not apparent in the data sequence, and so needs to be associated as a characteristic unit of the system.

Acquisition

Discrete-time signals may have several origins, but can usually be classified into one of two groups:

- By acquiring values of an analog signal at constant or variable rate. This process is called sampling.
- By observing an inherently discrete-time process, such as the weekly peak value of a particular economic indicator.

ESTIMATION THEORY

"Parameter estimation" redirects here. It is not to be confused with Point estimation or Interval estimation.

Estimation theory is a branch of statistics that deals with estimating the values of parameters based on measured empirical data that has a random component. The parameters describe an underlying physical setting in such a way that their value affects the distribution of the measured data. An estimator attempts to approximate the unknown parameters using the measurements.

For example, it is desired to estimate the proportion of a population of voters who will vote for a particular candidate. That proportion is the parameter sought; the estimate is based on a small random sample of voters.

Or, for example, in radar the goal is to estimate the range of objects (airplanes, boats, etc.) by analyzing the two-way transit timing of received echoes of transmitted pulses. Since the reflected pulses are unavoidably embedded in electrical noise, their measured values are randomly distributed, so that the transit time must be estimated.

In estimation theory, two approaches are generally considered.

- The probabilistic approach (described in this article) assumes that the measured data is random with probability distribution dependent on the parameters of interest
- The set-membership approach assumes that the measured data vector belongs to a set which depends on the parameter vector.

For example, in electrical communication theory, the measurements which contain information regarding the parameters of interest are often associated with a noisysignal.

Without randomness, or noise, the problem would be deterministic and estimation would not be needed.

Basics

To build a model, several statistical "ingredients" need to be known. These are needed to ensure the estimator has some mathematical tractability.

The first is a set of statistical samples taken from a random vector (RV) of size N. Put into a vector,

$$\mathbf{x} = \begin{bmatrix} x[0] \\ x[1] \\ \vdots \\ x[N-1] \end{bmatrix}$$

Secondly, there are the corresponding M parameters

$$\theta = \begin{bmatrix} \theta_1 \\ \theta_2 \\ \vdots \\ \theta_M \end{bmatrix},$$

which need to be established with their continuous probability density function (pdf) or its discrete counterpart, the probability mass function (pmf)

$$p(\mathbf{x}|\theta).$$

It is also possible for the parameters themselves to have a probability distribution (e.g., Bayesian statistics). It is then necessary to define the Bayesian probability

$$\pi(\theta).$$

After the model is formed, the goal is to estimate the parameters, commonly denoted $\hat{\theta},$, where the "hat" indicates the estimate.

One common estimator is the minimum mean squared error estimator, which utilizes the error between the estimated parameters and the actual value of the parameters

$$\mathbf{e} = \hat{\theta} - \theta$$

as the basis for optimality. This error term is then squared and minimized for the MMSE estimator.

Estimators

Commonly used estimators and estimation methods, and topics related to them:

- Maximum likelihood estimators
- Bayes estimators

- Method of moments estimators
- Cramér–Rao bound
- Minimum mean squared error (MMSE), also known as Bayes least squared error (BLSE)
- Maximum a posteriori (MAP)
- Minimum variance unbiased estimator (MVUE)
- nonlinear system identification
- Best linear unbiased estimator (BLUE)
- Unbiased estimators — see estimator bias.
- Particle filter
- Markov chain Monte Carlo (MCMC)
- Kalman filter, and its various derivatives
- Wiener filter

Examples

Unknown constant in additive white Gaussian noise

Consider a received discrete signal, $x[n]$, of N independent samples that consists of an unknown constant A with additive white Gaussian noise (AWGN) $w[n]$ with known variance σ^2 (*i.e.*, $\mathcal{N}(0,\sigma^2)$). Since the variance is known then the only unknown parameter is A. The model for the signal is then

$$x[n] = A + w[n] \quad n = 0, 1, \ldots, N-1$$

Two possible (of many) estimators for the parameter A are:

- $\hat{A}_1 = x[0]$
- $\hat{A}_2 = \frac{1}{N}\sum_{n=0}^{N-1} x[n]$ which is the sample mean

Both of these estimators have a mean of A, which can be shown through taking the expected value of each estimator

$$\mathrm{E}\left[\hat{A}_1\right] = \mathrm{E}\left[x[0]\right] = A$$

and

$$\mathrm{E}\left[\hat{A}_2\right] = \mathrm{E}\left[\frac{1}{N}\sum_{n=0}^{N-1} x[n]\right] = \frac{1}{N}\left[\sum_{n=0}^{N-1} \mathrm{E}\left[x[n]\right]\right] = \frac{1}{N}\left[NA\right] = A$$

At this point, these two estimators would appear to perform the same. However, the difference between them becomes apparent when comparing the variances.

$$\mathrm{var}\left(\hat{A}_1\right) = \mathrm{var}\left(x[0]\right) = \sigma^2$$

and

$$\text{var}\left(\hat{A}_2\right) = \text{var}\left(\frac{1}{N}\sum_{n=0}^{N-1} x[n]\right) \overset{\text{independence}}{=} \frac{1}{N^2}\left[\sum_{n=0}^{N-1} \text{var}(x[n])\right] = \frac{1}{N^2}\left[N\sigma^2\right] = \frac{\sigma^2}{N}$$

It would seem that the sample mean is a better estimator since its variance is lower for every N > 1.

Maximum Likelihood

Continuing the example using the maximum likelihood estimator, the probability density function (pdf) of the noise for one sample $w[n]$ is

$$p(w[n]) = \frac{1}{\sigma\sqrt{2\pi}} \exp\left(-\frac{1}{2\sigma^2} w[n]^2\right)$$

and the probability of $x[n]$ becomes $(x[n]$ can be thought of a $\mathcal{N}(A, \sigma^2)$

$$p(x[n]; A) = \frac{1}{\sigma\sqrt{2\pi}} \exp\left(-\frac{1}{2\sigma^2} (x[n] - A)^2\right)$$

By independence, the probability of x becomes

$$p(\mathbf{x}; A) = \prod_{n=0}^{N-1} p(x[n]; A) = \frac{1}{\left(\sigma\sqrt{2\pi}\right)^N} \exp\left(-\frac{1}{2\sigma^2}\sum_{n=0}^{N-1}(x[n] - A)^2\right)$$

Taking the natural logarithm of the pdf

$$\ln p(\mathbf{x}; A) = -N\ln\left(\sigma\sqrt{2\pi}\right) - \frac{1}{2\sigma^2}\sum_{n=0}^{N-1}(x[n] - A)^2$$

and the maximum likelihood estimator is

$$\hat{A} = \arg\max \ln p(\mathbf{x}; A)$$

Taking the first derivative of the log-likelihood function

$$\frac{\partial}{\partial A}\ln p(\mathbf{x}; A) = \frac{1}{\sigma^2}\left[\sum_{n=0}^{N-1}(x[n] - A)\right] = \frac{1}{\sigma^2}\left[\sum_{n=0}^{N-1} x[n] - NA\right]$$

and setting it to zero

$$0 = \frac{1}{\sigma^2}\left[\sum_{n=0}^{N-1} x[n] - NA\right] = \sum_{n=0}^{N-1} x[n] - NA$$

This results in the maximum likelihood estimator

$$\hat{A} = \frac{1}{N}\sum_{n=0}^{N-1} x[n]$$

which is simply the sample mean. From this example, it was found that the sample mean is the maximum likelihood estimator for N samples of a fixed, unknown parameter corrupted by AWGN.

Cramér–Rao Lower Bound

To find the Cramér–Rao lower bound (CRLB) of the sample mean estimator, it is first necessary to find the Fisher information number

$$\mathcal{I}(A) = \mathrm{E}\left(\left[\frac{\partial}{\partial A}\ln p(\mathbf{x};A)\right]^2\right) = -\mathrm{E}\left[\frac{\partial^2}{\partial A^2}\ln p(\mathbf{x};A)\right]$$

and copying from above

$$\frac{\partial}{\partial A}\ln p(\mathbf{x};A) = \frac{1}{\sigma^2}\left[\sum_{n=0}^{N-1} x[n] - NA\right]$$

Taking the second derivative

$$\frac{\partial^2}{\partial A^2}\ln p(\mathbf{x};A) = \frac{1}{\sigma^2}(-N) = \frac{-N}{\sigma^2}$$

and finding the negative expected value is trivial since it is now a deterministic constant

$$-\mathrm{E}\left[\frac{\partial^2}{\partial A^2}\ln p(\mathbf{x};A)\right] = \frac{N}{\sigma^2}$$

Finally, putting the Fisher information into

$$\mathrm{var}\left(\hat{A}\right) \geq \frac{1}{\mathcal{I}}$$

results in

$$\mathrm{var}\left(\hat{A}\right) \geq \frac{\sigma^2}{N}$$

Comparing this to the variance of the sample mean (determined previously) shows that the sample mean is equal to the Cramér–Rao lower bound for all values of N and A. In other words, the sample mean is the (necessarily unique) efficient estimator, and thus also the minimum variance unbiased estimator (MVUE), in addition to being the maximum likelihood estimator.

Maximum of a Uniform Distribution

One of the simplest non-trivial examples of estimation is the estimation of the maximum of a uniform distribution. It is used as a hands-on classroom exercise and to illustrate basic principles of estimation theory. Further, in the case of estimation based on a single sample, it demonstrates philosophical issues and possible misunderstandings in the use of maximum likelihood estimators and likelihood functions.

Given a discrete uniform distribution 1.2....N with unknown maximum, the UMVU estimator for the maximum is given by

$$\frac{k+1}{k}m - 1 = m + \frac{m}{k} - 1$$

where m is the sample maximum and k is the sample size, sampling without replacement. This problem is commonly known as the German tank problem, due to application of maximum estimation to estimates of German tank production during World War II.

The formula may be understood intuitively as:

The sample maximum plus the average gap between observations in the sample,

the gap being added to compensate for the negative bias of the sample maximum as an estimator for the population maximum.

This has a variance of

$$\frac{1}{k}\frac{(N-k)(N+1)}{(k+2)} \approx \frac{N^2}{k^2} \text{ for small samples } k \ll N$$

so a standard deviation of approximately N/k, the (population) average size of a gap between samples; compare $\frac{m}{k}$ above. This can be seen as a very simple case of maximum spacing estimation.

The sample maximum is the maximum likelihood estimator for the population maximum, but, as discussed above, it is biased.

Applications

Numerous fields require the use of estimation theory. Some of these fields include (but are by no means limited to):

- Interpretation of scientific experiments
- Signal processing
- Clinical trials
- Opinion polls
- Quality control
- Telecommunications
- Project management
- Software engineering
- Control theory (in particular Adaptive control)
- Network intrusion detection system
- Orbit determination

Measured data are likely to be subject to noise or uncertainty and it is through statistical probability that optimal solutions are sought to extract as much information from the data as possible.

NYQUIST–SHANNON SAMPLING THEOREM

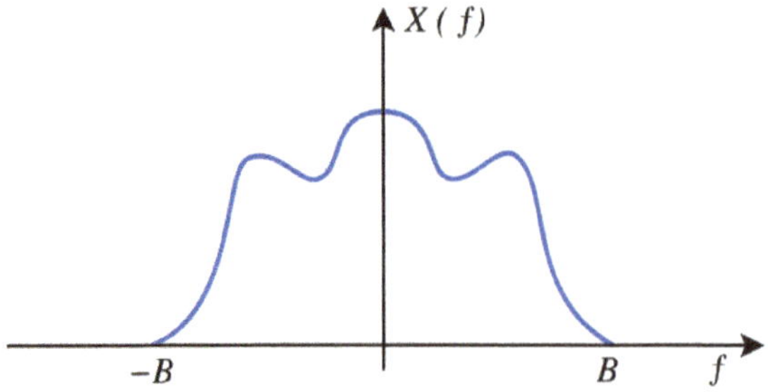

Example of magnitude of the Fourier transform of a bandlimited function

In the field of digital signal processing, the sampling theorem is a fundamental bridge between continuous-time signals (often called "analog signals") and discrete-time signals (often called "digital signals"). It establishes a sufficient condition for a sample rate that permits a discrete sequence of samples to capture all the information from a continuous-time signal of finite bandwidth.

Strictly speaking, the theorem only applies to a class of mathematical functions having a Fourier transform that is zero outside of a finite region of frequencies. Intuitively we expect that when one reduces a continuous function to a discrete sequence and interpolates back to a continuous function, the fidelity of the result depends on the density (or sample rate) of the original samples. The sampling theorem introduces the concept of a sample rate that is sufficient for perfect fidelity for the class of functions that are bandlimited to a given bandwidth, such that no actual information is lost in the sampling process. It expresses the sufficient sample rate in terms of the bandwidth for the class of functions. The theorem also leads to a formula for perfectly reconstructing the original continuous-time function from the samples.

Perfect reconstruction may still be possible when the sample-rate criterion is not satisfied, provided other constraints on the signal are known. (See § Sampling of non-baseband signals below, and compressed sensing.) In some cases (when the sample-rate criterion is not satisfied), utilizing additional constraints allows for approximate reconstructions. The fidelity of these reconstructions can be verified and quantified utilizing Bochner's theorem.

The name Nyquist–Shannon sampling theorem honors Harry Nyquist and Claude Shannon. The theorem was also discovered independently by E. T. Whittaker, by Vladimir Kotelnikov, and by others. It is thus also known by the names Nyquist–Shannon–Kotelnikov, Whittaker–Shannon–Kotelnikov, Whittaker–Nyquist–Kotelnikov–Shannon, and cardinal theorem of interpolation.

Sampling is a process of converting a signal (for example, a function of continuous time and/or space) into a numeric sequence (a function of discrete time and/or space). Shannon's version of the theorem states:

If a function x(t) contains no frequencies higher than B hertz, it is completely determined by giving its ordinates at a series of points spaced 1/(2B) seconds apart.

A sufficient sample-rate is therefore 2B samples/second, or anything larger. Equivalently, for a given sample rate fs, perfect reconstruction is guaranteed possible for a bandlimit B < fs/2.

When the bandlimit is too high (or there is no bandlimit), the reconstruction exhibits imperfections known as aliasing. Modern statements of the theorem are sometimes careful to explicitly state that x(t) must contain no sinusoidal component at exactly frequency B, or that B must be strictly less than ½ the sample rate. The two thresholds, 2B and fs/2 are respectively called the Nyquist rate and Nyquist frequency. And respectively, they are attributes of x(t) and of the sampling equipment. The condition described by these inequalities is called the Nyquist criterion, or sometimes the Raabe condition. The theorem is also applicable to functions of other domains, such as space, in the case of a digitized image. The only change, in the case of other domains, is the units of measure applied to t, fs, and B.

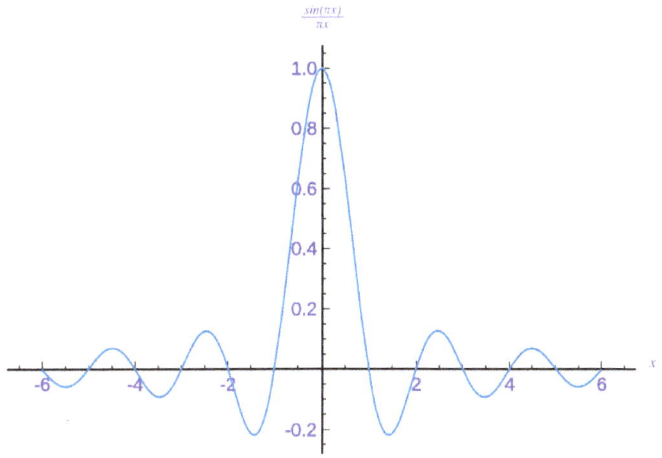

The normalized sinc function: sin(πx) / (πx) ... showing the central peak at x= 0, and zero-crossings at the other integer values of x.

The symbol T = 1/fs is customarily used to represent the interval between samples and is called the sample period or sampling interval. And the samples of function x(t) are commonly denoted by x[n] = x(nT) (alternatively "xn" in older signal processing literature), for all integer values of n. A mathematically ideal way to interpolate the sequence involves the use of sinc functions. Each sample in the sequence is replaced by a sinc function, centered on the time axis at the original location of the sample, nT, with the amplitude of the sinc function scaled to the sample value, x[n]. Subsequently, the sinc functions are summed into a continuous function. A mathematically equivalent method is to convolve one sinc function with a series of Dirac delta pulses, weighted by the sample values. Neither method is numerically practical. Instead, some type of approximation of the sinc functions, finite in length, is used. The imperfections attributable to the approximation are known as interpolation error.

Practical digital-to-analog converters produce neither scaled and delayed sinc functions, nor ideal Dirac pulses. Instead they produce a piecewise-constant sequence of scaled and delayed rectangular pulses (the zero-order hold), usually followed by an anti-imaging filter to clean up spurious high-frequency content.

Aliasing

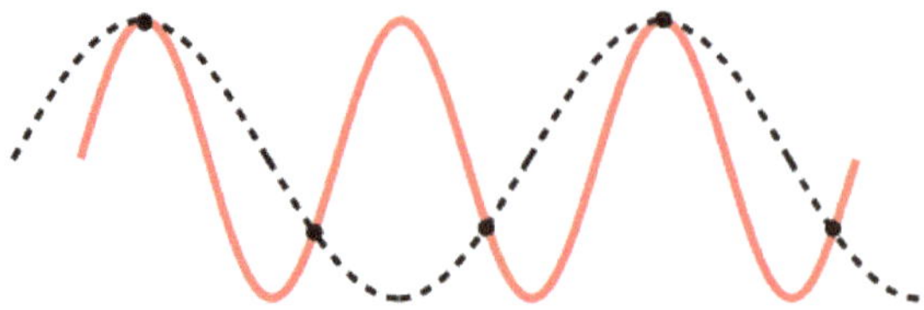

The samples of two sine waves can be identical when at least one of them is at a frequency above half the sample rate.

When $x(t)$ is a function with a Fourier transform, $X(f)$:

$$X(f) \stackrel{\text{def}}{=} \int_{-\infty}^{\infty} x(t)\, e^{-i2\pi ft}\, dt,$$

the Poisson summation formula indicates that the samples, x(nT), of x(t) are sufficient to create a periodic summation of X(f). The result is:

$$X_s(f) \stackrel{\text{def}}{=} \sum_{k=-\infty}^{\infty} X(f - kf_s) = \sum_{n=-\infty}^{\infty} T \cdot x(nT)\, e^{-i2\pi nTf},$$

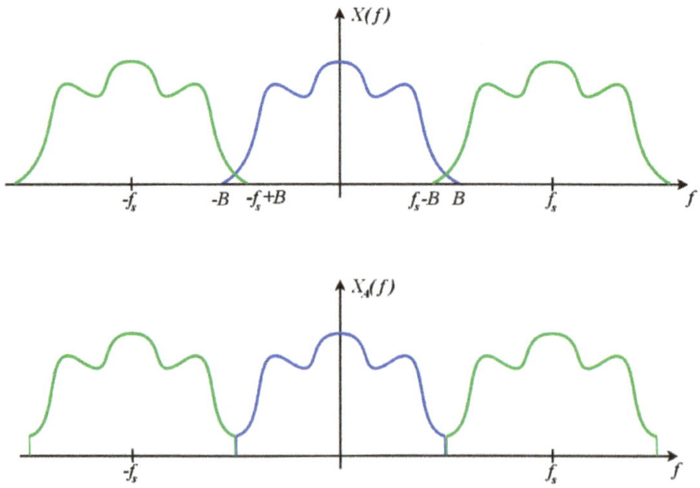

X(f) (top blue) and XA(f) (bottom blue) are continuous Fourier transforms of two different functions, x(t) and xA(t) (not shown). When the functions are sampled at rate fs, the images (green) are added to the original transforms (blue) when one examines the discrete-time Fourier transforms (DTFT) of the sequences. In this hypothetical example, the DTFTs are identical, which means the sampled sequences are identical, even though the original continuous pre-sampled functions are not. If these were audio signals, x(t) and xA(t) might not sound the same. But their samples (taken at rate fs) are identical and would lead to identical reproduced sounds; thus xA(t) is an alias of x(t) at this sample rate. which is a periodic function and its equivalent representation as a Fourier series, whose coefficients are T•x(nT). This function is also known as the discrete-time Fourier transform (DTFT) of the sequence T•x(nT), for integers n.

As depicted, copies of X(f) are shifted by multiples of fsand combined by addition. For a band-limited function (X(f) = 0 for all |f| ≥ B), and sufficiently large fs, it is possible for the copies to remain distinct from each other. But if the Nyquist criterion is not satisfied, adjacent copies overlap, and it is not possible in general to discern an unambiguous X(f). Any frequency component above fs/2 is indistinguishable from a lower-frequency component, called an alias, associated with one of the copies. In such cases, the customary interpolation techniques produce the alias, rather than the original component. When the sample-rate is pre-determined by other considerations (such as an industry standard), x(t) is usually filtered to reduce its high frequencies to acceptable levels before it is sampled. The type of filter required is a lowpass filter, and in this application it is called an anti-aliasing filter.

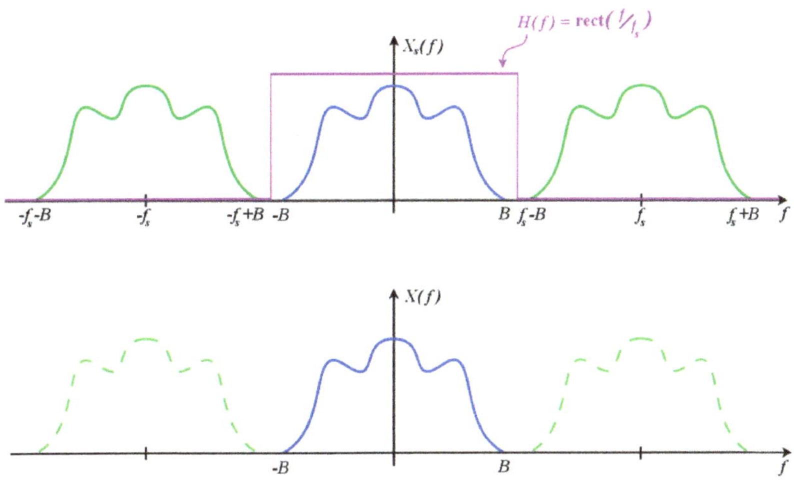

Spectrum, $X_s(f)$, of a properly sampled bandlimited signal (blue) and the adjacent DTFT images (green) that do not overlap. A *brick-wall* low-pass filter, $H(f)$, removes the images, leaves the original spectrum, $X(f)$, and recovers the original signal from its samples.

Derivation as a Special Case of Poisson Summation

When there is no overlap of the copies (aka "images") of X(f), the k = 0 term of Xs(f) can be recovered by the product:

$$X(f) = H(f) \cdot X_s(f), \quad \text{where:}$$

$$H(f) \stackrel{\text{def}}{=} \begin{cases} 1 & |f| < B \\ 0 & |f| > f_s - B. \end{cases}$$

At this point, the sampling theorem is proved, since $X(f)$ uniquely determines $x(t)$.

All that remains is to derive the formula for reconstruction. H(f) need not be precisely defined in the region [B, fs − B] because Xs(f) is zero in that region. However, the worst case is when B = fs/2, the Nyquist frequency. A function that is sufficient for that and all less severe cases is:

$$H(f) = \text{rect}\left(\frac{f}{f_s}\right) = \begin{cases} 1 & |f| < \frac{f_s}{2} \\ 0 & |f| > \frac{f_s}{2}, \end{cases}$$

where rect(•) is the rectangular function. Therefore:

$$X(f) = \text{rect}\left(\frac{f}{f_s}\right) \cdot X_s(f)$$

$$= \text{rect}(Tf) \cdot \sum_{n=-\infty}^{\infty} T \cdot x(nT)\, e^{-i2\pi nTf}$$

above).

$$= \sum_{n=-\infty}^{\infty} x(nT) \cdot \underbrace{T \cdot \text{rect}(Tf) \cdot e^{-i2\pi nTf}}_{\mathcal{F}\left\{\text{sinc}\left(\frac{t-nT}{T}\right)\right\}}.$$

The inverse transform of both sides produces the Whittaker–Shannon interpolation formula:

$$x(t) = \sum_{n=-\infty}^{\infty} x(nT) \cdot \text{sinc}\left(\frac{t - nT}{T}\right),$$

which shows how the samples, $x(nT)$, can be combined to reconstruct $x(t)$.

- Larger-than-necessary values of f_s (smaller values of T), called *oversampling*, have no effect on the outcome of the reconstruction and have the benefit of leaving room for a *transition band* in which $H(f)$ is free to take intermediate values. Undersampling, which causes aliasing, is not in general a reversible operation.
- Theoretically, the interpolation formula can be implemented as a low pass filter, whose impulse response is sinc(t/T) and whose input is $\sum_{n=-\infty}^{\infty} x(nT) \cdot \delta(t - nT)$, which is a Dirac comb function modulated by the signal samples. Practical digital-to-analog converters (DAC) implement an approximation like the zero-order hold. In that case, oversampling can reduce the approximation error.

Shannon's Original Proof

Poisson shows that the Fourier series in Eq.1 produces the periodic summation of X(f), regardless of fs and B. Shannon, however, only derives the series coefficients for the case fs = 2B. Virtually quoting Shannon's original paper:

Let $X(\omega)$ be the spectrum of $x(t)$ Then

$$\begin{aligned} x(t) &= \frac{1}{2\pi}\int_{-\infty}^{\infty} X(\omega)e^{i\omega t}\, d\omega \\ &= \frac{1}{2\pi}\int_{-2\pi B}^{2\pi B} X(\omega)e^{i\omega t}\, d\omega \end{aligned}$$

since $X(\omega)$ is assumed to be zero outside the band $|\frac{\omega}{2\pi}|<B.$. If we let

$$t = \frac{n}{2B}$$

where n is any positive or negative integer, we obtain

$$x\left(\frac{n}{2B}\right) = \frac{1}{2\pi}\int_{-2\pi B}^{2\pi B} X(\omega)e^{i\omega\frac{n}{2B}}\, d\omega.$$

On the left are values of $x(t)$ at the sampling points. The integral on the right will be recognized as essentially[n 1] the nthcoefficient in a Fourier-series expansion of the function n^{th} taking the interval –B to B as a fundamental period. This means that the values of the samples $x(n/2B)$ determine the Fourier coefficients in the series expansion of $X(\omega),$ Thus they determine $X(\omega),$ since $X(\omega),$ is zero for frequencies greater than B, and for lower frequencies $X(\omega),$ is determined if its Fourier coefficients are determined. But $X(\omega),$ determines the original function $x(t)$ completely, since a function is determined if its spectrum is known. Therefore the original samples determine the function $x(t)$ completely.

Shannon's proof of the theorem is complete at that point, but he goes on to discuss reconstruction via sinc functions, what we now call the Whittaker–Shannon interpolation formula as discussed above. He does not derive or prove the properties of the sinc function, but these would have been familiar to engineers reading his works at the time, since the Fourier pair relationship between rect (the rectangular function) and sinc was well known. Let x_n be the nth sample. Then the function $x(t)$ is represented by:

$$x(t) = \sum_{n=-\infty}^{\infty} x_n \frac{\sin \pi(2Bt - n)}{\pi(2Bt - n)}.$$

As in the other proof, the existence of the Fourier transform of the original signal is assumed, so the proof does not say whether the sampling theorem extends to bandlimited stationary random processes.

Application to Multivariable Signals and Images

Subsampled image showing a Moiré pattern

Properly sampled image

The sampling theorem is usually formulated for functions of a single variable. Consequently, the theorem is directly applicable to time-dependent signals and is normally formulated in that context. However, the sampling theorem can be extended in a straightforward way to functions of arbitrarily many variables. Grayscale images, for example, are often represented as two-dimensional arrays (or matrices) of real numbers representing the relative intensities of pixels (picture elements) located at the intersections of row and column sample locations. As a result, images require two independent variables, or indices, to specify each pixel uniquely—one for the row, and one for the column.

Color images typically consist of a composite of three separate grayscale images, one to represent each of the three primary colors—red, green, and blue, or RGB for short. Other colorspaces using 3-vectors for colors include HSV, CIELAB, XYZ, etc. Some colorspaces such as cyan, magenta, yellow, and black (CMYK) may represent color by four dimensions. All of these are treated as vector-valued functions over a two-dimensional sampled domain.

Similar to one-dimensional discrete-time signals, images can also suffer from aliasing if the sampling resolution, or pixel density, is inadequate. For example, a digital photograph of a striped shirt with high frequencies (in other words, the distance between the stripes is small), can cause aliasing of the shirt when it is sampled by the camera's image sensor. The aliasing appears as a moiré pattern. The solution to higher sampling in the spatial domain for this case would be to move closer to the shirt, use a higher resolution sensor, or to optically blur the image before acquiring it with the sensor.

Another example is shown to the right in the brick patterns. The top image shows the effects when the sampling theorem's condition is not satisfied. When software rescales an image (the same process that creates the thumbnail shown in the lower image) it, in effect, runs the image through a low-pass filter first and then downsamples the image to result in a smaller image that does not exhibit the moiré pattern. The top image is what happens when the image is downsampled without low-pass filtering: aliasing results.

The application of the sampling theorem to images should be made with care. For example, the sampling process in any standard image sensor (CCD or CMOS camera) is relatively far from the ideal sampling which would measure the image intensity at a single point. Instead these devices have a relatively large sensor area at each sample point in order to obtain sufficient amount of light. In other words, any detector has a finite-width point spread function. The analog optical image intensity function which is sampled by the sensor device is not in general bandlimited, and the non-ideal sampling is itself a useful type of low-pass filter, though not always sufficient to remove enough high frequencies to sufficiently reduce aliasing. When the area of the sampling spot (the size of the pixel sensor) is not large enough to provide sufficient spatial anti-aliasing, a separate anti-aliasing filter (optical low-pass filter) is typically included in a camera system to further blur the optical image. Despite images having these problems in relation to the sampling theorem, the theorem can be used to describe the basics of down and up sampling of images.

Critical Frequency

To illustrate the necessity of $fs > 2B$, consider the family of sinusoids generated by different values of θ in this formula:

$$x(t) = \frac{\cos(2\pi Bt + \theta)}{\cos(\theta)} = \cos(2\pi Bt) - \sin(2\pi Bt)\tan(\theta), \quad -\pi/2 < \theta < \pi/2.$$

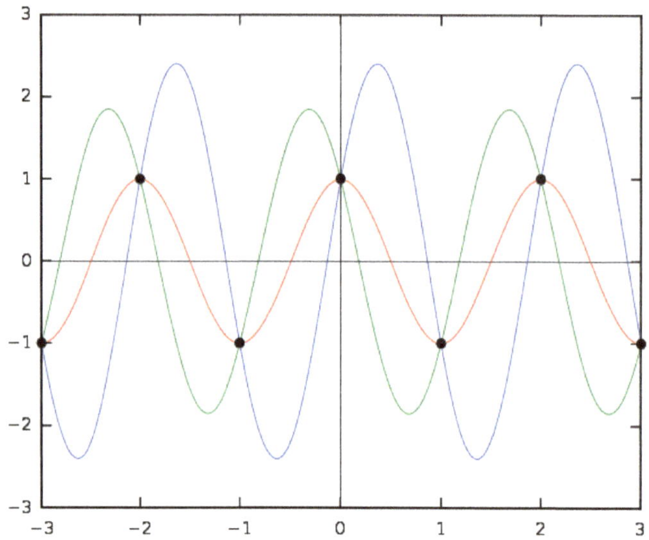

A family of sinusoids at the critical frequency, all having the same sample sequences of alternating +1 and –1. That is, they all are aliases of each other, even though their frequency is not above half the sample rate.

With fs = 2B or equivalently T = 1/(2B), the samples are given by:

$$x(nT) = \cos(\pi n) - \underbrace{\sin(\pi n)}_{0}\tan(\theta) = (-1)^n$$

regardless of the value of θ. That sort of ambiguity is the reason for the strict inequality of the sampling theorem's condition.

Sampling of non-baseband signals

As discussed by Shannon:

A similar result is true if the band does not start at zero frequency but at some higher value, and can be proved by a linear translation (corresponding physically to single-sideband modulation) of the zero-frequency case. In this case the elementary pulse is obtained from sin(x)/x by single-side-band modulation.

That is, a sufficient no-loss condition for sampling signals that do not have baseband components exists that involves the width of the non-zero frequency interval as opposed to its highest frequency component. See Sampling (signal processing) for more details and examples.

For example, in order to sample the FM radio signals in the frequency range of 100-102 MHz, it is not necessary to sample at 204 MHz (twice the upper frequency), but rather it is sufficient to sample at 4 MHz (twice the width of the frequency interval).

A bandpass condition is that X(f) = 0, for all nonnegative f outside the open band of frequencies:

$$\left(\frac{N}{2}f_s, \frac{N+1}{2}f_s\right)$$

for some nonnegative integer N. This formulation includes the normal baseband condition as the case N=0.

The corresponding interpolation function is the impulse response of an ideal brick-wall bandpass filter (as opposed to the ideal brick-wall lowpass filter used above) with cutoffs at the upper and lower edges of the specified band, which is the difference between a pair of lowpass impulse responses:

$$(N+1)\ \mathrm{sinc}\left(\frac{(N+1)t}{T}\right) - N\ \mathrm{sinc}\left(\frac{Nt}{T}\right)$$

Other generalizations, for example to signals occupying multiple non-contiguous bands, are possible as well. Even the most generalized form of the sampling theorem does not have a provably true converse. That is, one cannot conclude that information is necessarily lost just because the conditions of the sampling theorem are not satisfied; from an engineering perspective, however, it is generally safe to assume that if the sampling theorem is not satisfied then information will most likely be lost.

Nonuniform Sampling

The sampling theory of Shannon can be generalized for the case of nonuniform sampling, that is, samples not taken equally spaced in time. The Shannon sampling theory for non-uniform sampling states that a band-limited signal can be perfectly reconstructed from its samples if the average sampling rate satisfies the Nyquist condition. Therefore, although uniformly spaced samples may result in easier reconstruction algorithms, it is not a necessary condition for perfect reconstruction.

The general theory for non-baseband and nonuniform samples was developed in 1967 by Landau. He proved that the average sampling rate (uniform or otherwise) must be twice the occupied bandwidth of the signal, assuming it is a priori known what portion of the spectrum was occupied. In the late 1990s, this work was partially extended to cover signals of when the amount of occupied bandwidth was known, but the actual occupied portion of the spectrum was unknown. In the 2000s, a complete theory was developed (see the section Beyond Nyquist below) using compressed sensing. In particular, the theory, using signal processing language, is described in this 2009 paper. They show, among other things, that if the frequency locations are unknown, then it is necessary to sample at least at twice the Nyquist criteria; in other words, you must pay at least a factor of 2 for not knowing the location of the spectrum. Note that minimum sampling requirements do not necessarily guarantee stability.

Sampling Below the Nyquist Rate under Additional Restrictions

The Nyquist–Shannon sampling theorem provides a sufficient condition for the sampling and reconstruction of a band-limited signal. When reconstruction is done via the Whittaker–

Shannon interpolation formula, the Nyquist criterion is also a necessary condition to avoid aliasing, in the sense that if samples are taken at a slower rate than twice the band limit, then there are some signals that will not be correctly reconstructed. However, if further restrictions are imposed on the signal, then the Nyquist criterion may no longer be a necessary condition.

A non-trivial example of exploiting extra assumptions about the signal is given by the recent field of compressed sensing, which allows for full reconstruction with a sub-Nyquist sampling rate. Specifically, this applies to signals that are sparse (or compressible) in some domain. As an example, compressed sensing deals with signals that may have a low over-all bandwidth (say, the effective bandwidth EB), but the frequency locations are unknown, rather than all together in a single band, so that the passband technique doesn't apply. In other words, the frequency spectrum is sparse. Traditionally, the necessary sampling rate is thus 2B. Using compressed sensing techniques, the signal could be perfectly reconstructed if it is sampled at a rate slightly lower than 2EB. The downside of this approach is that reconstruction is no longer given by a formula, but instead by the solution to a convex optimization program which requires well-studied but nonlinear methods.

Historical Background

The sampling theorem was implied by the work of Harry Nyquist in 1928, in which he showed that up to 2B independent pulse samples could be sent through a system of bandwidth B; but he did not explicitly consider the problem of sampling and reconstruction of continuous signals. About the same time, Karl Küpfmüller showed a similar result, and discussed the sinc-function impulse response of a band-limiting filter, via its integral, the step response Integralsinus; this bandlimiting and reconstruction filter that is so central to the sampling theorem is sometimes referred to as a Küpfmüller filter (but seldom so in English).

The sampling theorem, essentially a dual of Nyquist's result, was proved by Claude E. Shannon. V. A. Kotelnikov published similar results in 1933, as did the mathematician E. T. Whittaker in 1915, J. M. Whittaker in 1935, and Gabor in 1946 ("Theory of communication"). In 1999, the Eduard Rhein Foundation awarded Kotelnikov their Basic Research Award for the first theoretically exact formulation of the sampling theorem.

In 1948 and 1949, Claude E. Shannon published the two revolutionary papers in which he founded the information theory. In Shannon 1948 the sampling theorem is formulated as "Theorem 13": Let f(t) contain no frequencies over W. Then

$$f(t) = \sum_{n=-\infty}^{\infty} X_n \frac{\sin \pi(2Wt - n)}{\pi(2Wt - n)},$$

where Xn = f(n/2W). It was not until these papers were published that the theorem known as "Shannon's sampling theorem" became common property among communication engineers, although Shannon himself writes that this is a fact which is common knowledge in the communication art.[note 2] A few lines further on, however, he adds: ... "but in spite of its evident importance [it] seems not to have appeared explicitly in the literature of communication theory.

Other Discoverers

Others who have independently discovered or played roles in the development of the sampling theorem have been discussed in several historical articles, for example by Jerri and by Lüke. For example, Lüke points out that H. Raabe, an assistant to Küpfmüller, proved the theorem in his 1939 Ph.D. dissertation; the term Raabe condition came to be associated with the criterion for unambiguous representation (sampling rate greater than twice the bandwidth). Meijering mentions several other discoverers and names in a paragraph and pair of footnotes:

As pointed out by Higgins , the sampling theorem should really be considered in two parts, as done above: the first stating the fact that a bandlimited function is completely determined by its samples, the second describing how to reconstruct the function using its samples. Both parts of the sampling theorem were given in a somewhat different form by J. M. Whittaker [350, 351, 353] and before him also by Ogura [241, 242]. They were probably not aware of the fact that the first part of the theorem had been stated as early as 1897 by Borel .27 As we have seen, Borel also used around that time what became known as the cardinal series. However, he appears not to have made the link . In later years it became known that the sampling theorem had been presented before Shannon to the Russian communication community by Kotel'nikov . In more implicit, verbal form, it had also been described in the German literature by Raabe . Several authors [33, 205] have mentioned that Someya introduced the theorem in the Japanese literature parallel to Shannon. In the English literature, Weston introduced it independently of Shannon around the same time.

Several authors, following Black , have claimed that this first part of the sampling theorem was stated even earlier by Cauchy, in a paper published in 1841. However, the paper of Cauchy does not contain such a statement, as has been pointed out by Higgins .

As a consequence of the discovery of the several independent introductions of the sampling theorem, people started to refer to the theorem by including the names of the aforementioned authors, resulting in such catchphrases as "the Whittaker–Kotel'nikov–Shannon (WKS) sampling theorem or even the Whittaker–Kotel'nikov–Raabe–Shannon–Someya sampling theorem . To avoid confusion, perhaps the best thing to do is to refer to it as the sampling theorem, rather than trying to find a title that does justice to all claimants .\

Why Nyquist?

Exactly how, when, or why Harry Nyquist had his name attached to the sampling theorem remains obscure. The term Nyquist Sampling Theorem (capitalized thus) appeared as early as 1959 in a book from his former employer, Bell Labs, and appeared again in 1963, and not capitalized in 1965. It had been called the Shannon Sampling Theorem as early as 1954, but also just the sampling theorem by several other books in the early 1950s.

In 1958, Blackman and Tukey cited Nyquist's 1928 paper as a reference for *the sampling theorem of information theory*,even though that paper does not treat sampling and reconstruction of continuous signals as others did. Their glossary of terms includes these entries:

Sampling theorem (of information theory)

Nyquist's result that equi-spaced data, with two or more points per cycle of highest frequency, allows reconstruction of band-limited functions. (See *Cardinal theorem.*)

Cardinal theorem (of interpolation theory)

A precise statement of the conditions under which values given at a doubly infinite set of equally spaced points can be interpolated to yield a continuous band-limited function with the aid of the function

$$\frac{\sin(x - x_i)}{x - x_i}.$$

Exactly what "Nyquist's result" they are referring to remains mysterious.

When Shannon stated and proved the sampling theorem in his 1949 paper, according to Meijering "he referred to the critical sampling interval T = 1/(2W) as the Nyquist interval corresponding to the band W, in recognition of Nyquist's discovery of the fundamental importance of this interval in connection with telegraphy." This explains Nyquist's name on the critical interval, but not on the theorem.

Similarly, Nyquist's name was attached to Nyquist rate in 1953 by Harold S. Black:

If the essential frequency range is limited to B cycles per second, 2B was given by Nyquist as the maximum number of code elements per second that could be unambiguously resolved, assuming the peak interference is less half a quantum step. This rate is generally referred to as signaling at the Nyquist rate and 1/(2B) has been termed a Nyquist interval." (bold added for emphasis; italics as in the original)

According to the OED, this may be the origin of the term *Nyquist rate*. In Black's usage, it is not a sampling rate, but a signaling rate.

REFERENCES

1. Nemirovsky, Jonathan; Shimron, Efrat (2015). Utilizing Bochners Theorem for Constrained Evaluation of Missing Fourier Data. arXiv:1506.03300 [physics.med-ph].
2. Shannon, Claude E. (January 1949). Communication in the presence of noise. Proc. Institute of Radio Engineers. 37 *(1): 10–21.* Reprint as classic paper in: Proc. IEEE, Vol. 86, No. 2, (Feb 1998)
3. Marvasti (ed), F. (2000). Nonuniform Sampling, Theory and Practice. New York: Kluwer Academic/Plenum Publishers.
4. Landau, H. J. (1967). Necessary density conditions for sampling and interpolation of certain entire functions. Acta Math. 117 *(1): 37–52.* doi:10.1007/BF02395039.
5. see, e.g., Feng, P. (1997). Universal minimum-rate sampling and spectrum-blind reconstruction for multiband signals. Ph.D. dissertation, University of Illinois at Urbana-Champaign.
6. Mishali, Moshe; Eldar, Yonina C. (March 2009). Blind Multiband Signal Reconstruction: Compressed Sensing for Analog Signals. IEEE Trans. Signal Processing. 57 *(3).* CiteSeerX *10.1.1.154.4255.*
7. Nyquist, Harry (April 1928). Certain topics in telegraph transmission theory. Trans. AIEE. 47: 617–644. Reprint as classic paper in:Proc. IEEE, Vol. 90, No. 2, Feb 2002

8. Küpfmüller, Karl (1928). *Über* die Dynamik der selbsttätigen Verstärkungsregler. Elektrische Nachrichtentechnik *(in German). 5 (11): 459–467.* (English translation 2005).

9. Kotelnikov, V.A. (1933). On the carrying capacity of the ether and wire in telecommunications. Material for the First All-Union Conference on Questions of Communication, Izd. Red. Upr. Svyazi RKKA (in Russian). Moscow. (English translation, PDF)

10. Whittaker, E.T. (1915). On the Functions Which are Represented by the Expansions of the Interpolation Theory. Proc. Royal Soc. Edinburgh. 35: 181–194. ("Theorie der Kardinalfunktionen).

11. Whittaker, J.M. (1935). Interpolatory Function Theory. Cambridge, England: Cambridge Univ. Press..

12. Shannon, Claude E. (July 1948). A Mathematical Theory of Communication. Bell System Technical Journal. 27 *(3): 379–423. doi:10.1002/j.1538-7305.1948.tb01338.x..*

13. Shannon, Claude E. (October 1948). A Mathematical Theory of Communication. Bell System Technical Journal. 27 *(4): 623–666. doi:10.1002/j.1538-7305.1948.tb00917.x.*

14. Jerri, Abdul (November 1977). The Shannon Sampling Theorem—Its Various Extensions and Applications: A Tutorial Review. Proceedings of the IEEE. 65 *(11): 1565–1596.* See also Jerri, Abdul (April 1979). Correction to The Shannon sampling theorem—Its various extensions and applications: A tutorial review. Proceedings of the IEEE. 67 *(4): 695.*

15. Lüke, Hans Dieter (April 1999). The Origins of the Sampling Theorem. IEEE Communications Magazine (4): 106–108. doi:10.1109/35.755459.

16. Meijering, Erik (March 2002). A Chronology of Interpolation From Ancient Astronomy to Modern Signal and Image Processing. Proc. IEEE. 90 *(3): 319–322. doi:10.1109/5.993400.*

17. Members of the Technical Staff of Bell Telephone Lababoratories (1959). Transmission Systems for Communications. AT&T. pp. 26–4 (Vol.2).

18. Guillemin, Ernst Adolph (1963). Theory of Linear Physical Systems. Wiley.

19. Roberts, Richard A.; Barton, Ben F. (1965). Theory of Signal Detectability: Composite Deferred Decision Theory.

20. Gray, Truman S. (1954). Applied Electronics: A First Course in Electronics, Electron Tubes, and Associated Circuits.

21. Blackman, R. B.; Tukey, J. W. (1958). The Measurement of Power Spectra : From the Point of View of Communications Engineering (PDF). New York: Dover.

22. Black, Harold S. (1953). Modulation Theory.

CHAPTER 3

Sub-fields of Digital Signal Processing

Chapter Contents

- **Audio Signal Processing**
- **History**
- **Digital Signals**
- **Application Areas**
- **Digital Image Processing**
- **Speech Processing**
- **Statistical Signal Processing**

In the sense of hearing provides us rich information about our environment with respect to the locations and characteristics of sound producing objects. For example, we can effortlessly assimilate the sounds of birds twittering outside the window and traffic moving in the distance while following the lyrics of a song over the radio sung with multi-instrument accompaniment. The human auditory system is able to process the complex sound mixture reaching our ears and form high-level abstractions of the environment by the analysis and grouping of measured sensory inputs. The process of achieving the segregation and identification of sources from the received composite acoustic signal is known as auditory scene analysis. It is easy to imagine that the machine realization of this functionality (sound source separation and classification) would be very useful in applications such as speech recognition in noise, automatic music transcription and multimedia data search and retrieval. In all cases the audio signal must be processed based on signal models, which may be drawn from sound production as well as sound perception and cognition. While production models are an integral part of speech processing systems, general audio processing is still limited to rather basic signal models due to the diverse and wide-ranging nature of audio signals.

AUDIO SIGNAL PROCESSING

Audio signal processing or audio processing is the intentional alteration of audio signals often through an audio effect or effects unit. As audio signals may be electronically represented in either digital or analog format, signal processing may occur in either domain. Analog processors operate directly on the electrical signal, while digital processors operate mathematically on the digital representation of that signal.

History

Audio signals are electronic representations of sound waves—longitudinal waves which travel through air, consisting of compressions and rarefactions. The energy contained in audio signals is typically measured in decibels. Audio processing was necessary for early radio broadcasting, as there were many problems with studio to transmitter links.[1]

Analog Signals

"Analog" indicates something that is mathematically represented by a set of continuous values; for example, the analog clock uses constantly moving hands on a physical clock face, where moving the hands directly alters the information that clock is providing. Thus, an analog signal is one represented by a continuous stream of data, in this case along an electrical circuit in the form of voltage, current or charge changes (compare with digital signals below). Analog signal processing (ASP) then involves physically altering the continuous signal by changing the voltage or current or charge via various electrical means.

Historically, before the advent of widespread digital technology, ASP was the only method by which to manipulate a signal. Since that time, as computers and software became more advanced, digital signal processing has become the method of choice.

DIGITAL SIGNALS

A digital representation expresses the pressure wave-form as a sequence of symbols, usually binary numbers. This permits signal processing using digital circuits such as microprocessors and computers. Although such a conversion can be prone to loss, most modern audio systems use this approach as the techniques of digital signal processing are much more powerful and efficient than analog domain signal processing.[2]

APPLICATION AREAS

Processing methods and application areas include storage, level compression, data compression, transmission, enhancement (e.g., equalization, filtering, noise cancellation, echo or reverb removal or addition, etc.)

Audio Broadcasting

Traditionally the most important audio processing (in audio broadcasting) takes place just before the transmitter. Studio audio processing is limited in the modern era due to digital audio systems (mixers, routers) being pervasive in the studio.

In audio broadcasting, the audio processor must

- prevent or minimize overmodulation,
- compensate for non-linear transmitters (a potential issue with medium wave and shortwave broadcasting) and
- adjust overall loudness to desired level

Techniques

Audio unprocessed by reverb and delay is metaphorically referred to as "dry", while processed audio is referred to as "wet".[3]

- *echo* - to simulate the effect of reverberation in a large hall or cavern, one or several delayed signals are added to the original signal. To be perceived as echo, the delay has to be of order 35 milliseconds or above. Short of actually playing a sound in the desired environment, the effect of echo can be implemented using either digital or analog methods. Analog echo effects are implemented using tape delays and/or spring reverbs. When large numbers of delayed signals are mixed over several seconds, the resulting sound has the effect of being presented in a large room, and it is more commonly called reverberation or reverb for short.
- *flanger* - to create an unusual sound, a delayed signal is added to the original signal with a continuously variable delay (usually smaller than 10 ms). This effect is now done electronically using DSP, but originally the effect was created by playing the same recording on two synchronized tape players, and then mixing the signals together. As long as the machines were synchronized, the mix would sound more-or-less normal, but if the operator placed his finger on the flange of one of the players (hence flanger), that machine would slow down and its signal would fall out-of-phase with its partner, producing a phasing effect. Once the operator took his finger off, the player would speed up until its tachometer was back in phase with the master, and as this happened, the phasing effect would appear to slide up the frequency spectrum. This phasing up-and-down the register can be performed rhythmically.
- *phaser* - another way of creating an unusual sound; the signal is split, a portion is filtered with an all-pass filter to produce a phase-shift, and then the unfiltered and filtered signals are mixed. The phaser effect was originally a simpler implementation of the flanger effect since delays were difficult to implement with analog equipment. Phasers are often used to give a synthesized or electronic effect to natural sounds, such as human speech. The voice of C-3PO from Star Wars was created by taking the actor's voice and treating it with a phaser.
- *chorus* - a delayed signal is added to the original signal with a constant delay. The delay has to be short in order not to be perceived as echo, but above 5 ms to be audible. If the delay is too short, it will destructively interfere with the un-delayed

signal and create a flanging effect. Often, the delayed signals will be slightly pitch shifted to more realistically convey the effect of multiple voices.

- *equalization* - different frequency bands are attenuated or boosted to produce desired spectral characteristics. Moderate use of equalization (often abbreviated as "EQ") can be used to "fine-tune" the tone quality of a recording; extreme use of equalization, such as heavily cutting a certain frequency can create more unusual effects.
- *filtering* - Equalization is a form of filtering. In the general sense, frequency ranges can be emphasized or attenuated using low-pass, high-pass, band-pass or band-stop filters. Band-pass filtering of voice can simulate the effect of a telephone because telephones use band-pass filters.
- *overdrive* effects such as the use of a fuzz box can be used to produce distorted sounds, such as for imitating robotic voices or to simulate distorted radiotelephone traffic (e.g., the radio chatter between starfighter pilots in the science fiction film *Star Wars*). The most basic overdrive effect involves *clipping* the signal when its absolute value exceeds a certain threshold.
- *pitch shift* - this effect shifts a signal up or down in pitch. For example, a signal may be shifted an octave up or down. This is usually applied to the entire signal, and not to each note separately. Blending the original signal with shifted duplicate(s) can create harmonies from one voice. Another application of pitch shifting is pitch correction. Here a musical signal is tuned to the correct pitch using digital signal processing techniques. This effect is ubiquitous in karaoke machines and is often used to assist pop singers who sing out of tune. It is also used intentionally for aesthetic effect in such pop songs as Cher's *Believe* and Madonna's *Die Another Day*.
- *time stretching* - the complement of pitch shift, that is, the process of changing the speed of an audio signal without affecting its pitch.
- *resonators* - emphasize harmonic frequency content on specified frequencies. These may be created from parametric EQs or from delay-based comb-filters.
- *robotic voice effects* are used to make an actor's voice sound like a synthesized human voice.
- *synthesizer* - generate artificially almost any sound by either imitating natural sounds or creating completely new sounds.
- *modulation* - to change the frequency or amplitude of a carrier signal in relation to a predefined signal. Ring modulation, also known as amplitude modulation, is an effect made famous by Doctor Who's Daleks and commonly used throughout sci-fi.
- *compression* - the reduction of the dynamic range of a sound to avoid unintentional fluctuation in the dynamics. Level compression is not to be confused with audio data compression, where the amount of data is reduced without affecting the amplitude of the sound it represents.
- *3D audio effects* - place sounds outside the stereo basis
- *reverse echo* - a swelling effect created by reversing an audio signal and recording echo and/or delay while the signal runs in reverse. When played back forward the last echos are heard before the effected sound creating a rush like swell preceding

and during playback. Jimmy Page of Led Zeppelin used this effect in the bridge of "Whole Lotta Love.

- *active noise control*- a method for reducing unwanted sound
- *wave field synthesis* - a spatial audio rendering technique for the creation of virtual acoustic environments

DIGITAL IMAGE PROCESSING

Digital image processing is the use of computer algorithms to perform image processing on digital images. As a subcategory or field of digital signal processing, digital image processing has many advantages over analog image processing. It allows a much wider range of algorithms to be applied to the input data and can avoid problems such as the build-up of noise and signal distortion during processing. Since images are defined over two dimensions (perhaps more) digital image processing may be modeled in the form of multidimensional systems.

Many of the techniques of digital image processing, or digital picture processing as it often was called, were developed in the 1960s at the Jet Propulsion Laboratory, Massachusetts Institute of Technology, Bell Laboratories, University of Maryland, and a few other research facilities, with application to satellite imagery, wire-photo standards conversion, medical imaging, videophone, character recognition, and photograph enhancement.[1] The cost of processing was fairly high, however, with the computing equipment of that era. That changed in the 1970s, when digital image processing proliferated as cheaper computers and dedicated hardware became available. Images then could be processed in real time, for some dedicated problems such as television standards conversion. As general-purpose computers became faster, they started to take over the role of dedicated hardware for all but the most specialized and computer-intensive operations.

With the fast computers and signal processors available in the 2000s, digital image processing has become the most common form of image processing and generally, is used because it is not only the most versatile method, but also the cheapest.

Digital image processing technology for medical applications was inducted into the Space Foundation Space Technology Hall of Fame in 1994.[2]

In 2002 Raanan Fattel, introduced Gradient domain image processing, a new way to process images in which the differences between pixels are manipulated rather than the pixel values themselves.

Tasks

Digital image processing allows the use of much more complex algorithms, and hence, can offer both more sophisticated performance at simple tasks, and the implementation of methods which would be impossible by analog means.

In particular, digital image processing is the only practical technology for:

- Classification
- Feature extraction

- Multi-scale signal analysis
- Pattern recognition
- Projection

Some techniques which are used in digital image processing include:

- Anisotropic diffusion
- Hidden Markov models
- Image editing
- Image restoration
- Independent component analysis
- Linear filtering
- Neural networks
- Partial differential equations
- Pixelation
- Principal components analysis
- Self-organizing maps
- Wavelets

Applications

Digital Camera Images

Digital cameras generally include specialized digital image processing hardware – either dedicated chips or added circuitry on other chips – to convert the raw data from their image sensor into a color-corrected image in a standard image file format

Film

Westworld (1973) was the first feature film to use the digital image processing to pixellate photography to simulate an android's point of view.

SPEECH PROCESSING

Speech processing is the study of speech signals and the processing methods of these signals. The signals are usually processed in a digital representation, so speech processing can be regarded as a special case of digital signal processing, applied to speech signal. Aspects of speech processing includes the acquisition, manipulation, storage, transfer and output of speech signals. The input is called speech recognition and the output is called speech synthesis.

STATISTICAL SIGNAL PROCESSING

Statistical signal processing is an area of Applied Mathematics and Signal Processing that treats signals as stochastic processes, dealing with their statistical properties (e.g., mean, covariance, etc.). Because of its very broad range of application Statistical signal processing is taught at the graduate level in either Electrical Engineering, Applied Mathematics, Pure Mathematics/Statistics, or even Biomedical Engineering and Physics departments around the world, although important applications exist in almost all scientific fields.

Basic Signal Model

In many applications, a signal is modeled as functions consisting of both a deterministic and a stochastic component. A simple example and also a common model of many statistical systems is a signal $y(t)$ that consists of a deterministic part $x(t)$ added to noise which can be modeled in many situations as white Gaussian noise $w(t)$

$$y(t) = x(t) + w(t)$$

where

$$w(t) \sim \mathcal{N}(0, \sigma^2)$$

White noise simply means that the noise process is completely uncorrelated. As a result, its autocorrelation function is an impulse:

$$R_{ww}(\tau) = \sigma^2 \delta(\tau)$$

where

$\delta(\tau)$ is the Dirac delta function.

Given information about a statistical system and the random variable from which it is derived, we can increase our knowledge of the output signal; conversely, given the statistical properties of the output signal, we can infer the properties of the underlying random variable. These statistical techniques are developed in the fields of estimation theory, detection theory, and numerous related fields that rely on statistical information to maximize their efficiency.

Example

The Computation of Average Transients (CAT) is used routinely in FT-NMR spectroscopy (nuclear magnetic resonance) to improve the signal-noise ratio of nmr spectra. The signal is measured repeatedly n times and then averaged.

$$\bar{y} = \frac{1}{n} \sum_i y(t)_i = x(t) + \frac{1}{n} \sum_i w(t)_i$$

Assuming that the noise is white and that its variance is constant in time it follows by error propagation that

$$\sigma(\bar{y}) = \frac{1}{\sqrt{n}}\sigma$$

Thus, if 10,000 measurements are averaged the signal to noise ratio is increased by a factor of 100, enabling the measurement of ^{13}C NMR spectra at natural abundance (1.1%) of ^{13}C.

REFERENCES

1. Atti, Andreas Spanias, Ted Painter, Venkatraman (2006). Audio signal processing and coding ([Online-Ausg.] ed.). Hoboken, NJ: John Wiley & Sons. p. 464. ISBN 0-471-79147-4.
2. Zölzer, Udo (1997). Digital Audio Signal Processing. John Wiley and Sons. ISBN 0-471-97226-6.
3. Hodgson, Jay (2010). Understanding Records, p.95. ISBN 978-1-4411-5607-5.
4. Azriel Rosenfeld, Picture Processing by Computer, New York: Academic Press, 1969
5. Space Technology Hall of Fame:Inducted Technologies/1994. Space Foundation. 1994. Archived from the original on 4 July 2011. Retrieved 7 January 2010.
6. Bhat, Pravin, et al. Gradientshop: A gradient-domain optimization framework for image and video filtering. ACM Transactions on Graphics 29.2 (2010): 10.
7. A Brief, Early History of Computer Graphics in Film Archived 17 July 2012 at the Wayback Machine., Larry Yaeger, 16 August 2002 (last update), retrieved 24 March 2010

CHAPTER 4

Advanced Z-transform

Chapter Contents

- **Introduction**
- **Properties**
- **Bilinear Transform**
- **Discrete Fourier Transform**
- **Discrete-Time Fourier Transform**
- **Impulse Invariance**
- **Matched Z-Transform Method**
- **Z-Transform**
- **Zak Transform**

The unilateral z-transform of an arbitrary signal $x[n]$ is defined as

$$\mathcal{UZ}[x[n]] = X(z) \stackrel{\triangle}{=} \sum_{n=-\infty}^{\infty} x[n]u[n]z^{-n} = \sum_{n=0}^{\infty} x[n]z^{-n}$$

When the unilateral z-transform is applied to find the transfer function $H(z) = \mathcal{UZ}[h[n]]$ of an LTI system, it is always assumed to be causal, and the ROC is always the exterior of a circle. The unilateral z-transform of any signal $x[n] = x[n]u[n]$ is identical to its bilateral Laplace transform. However, if $x[n] \neq x[n]u[n]$, the two z-transforms are different. Some of the properties of the unilateral z-transform different from the bilateral z-transform are listed below.

INTRODUCTION

In mathematics and signal processing, the advanced Z-transform is an extension of the Z-transform, to incorporate ideal delays that are not multiples of the sampling time. It takes the form

$$F(z,m)=\sum_{k=0}^{\infty} f(kT+m)z^{-k}$$

where

- T is the sampling period
- m (the "delay parameter") is a fraction of the sampling period {\displaystyle [0,T].}

It is also known as the **modified Z-transform**.

The advanced Z-transform is widely applied, for example to accurately model processing delays in digital control.

PROPERTIES

If the delay parameter, m, is considered fixed then all the properties of the Z-transform hold for the advanced Z-transform.

Linearity

$$\mathcal{Z}\left\{\sum_{k=1}^{n} c_k f_k(t)\right\}=\sum_{k=1}^{n} c_k F(z,m).$$

Time shift

$$\mathcal{Z}\left\{u(t-nT)f(t-nT)\right\}=z^{-n}F(z,m).$$

Damping

$$\mathcal{Z}\left\{f(t)e^{-a\,t}\right\}=e^{-a\,m}F(e^{a\,T}z,m).$$

Time multiplication

$$\mathcal{Z}\left\{t^y f(t)\right\}=\left(-Tz\frac{d}{dz}+m\right)^y F(z,m).$$

Final value theorem

$$\lim_{k\to\infty} f(kT+m)=\lim_{z\to 1}(1-z^{-1})F(z,m).$$

Example

Consider the following example where {\displaystyle f(t)=\cos(\omega t)}:

$$\begin{aligned}F(z,m) &= \mathcal{Z}\{\cos(\omega(kT+m))\}\\ &= \mathcal{Z}\{\cos(\omega kT)\cos(\omega m) - \sin(\omega kT)\sin(\omega m)\}\\ &= \cos(\omega m)\mathcal{Z}\{\cos(\omega kT)\} - \sin(\omega m)\mathcal{Z}\{\sin(\omega kT)\}\\ &= \cos(\omega m)\frac{z(z-\cos(\omega T))}{z^2-2z\cos(\omega T)+1} - \sin(\omega m)\frac{z\sin(\omega T)}{z^2-2z\cos(\omega T)+1}\\ &= \frac{z^2\cos(\omega m) - z\cos(\omega(T-m))}{z^2-2z\cos(\omega T)+1}.\end{aligned}$$

If $m = 0$ then $F(z,m)$ reduces to the transform

$$F(z,0) = \frac{z^2 - z\cos(\omega T)}{z^2-2z\cos(\omega T)+1},$$

which is clearly just the Z-transform of f(t).

BILINEAR TRANSFORM

The Bilinear transform (also known as Tustin's method or Mobius Transformation) is used in digital signal processing and discrete-time control theory to transform continuous-time system representations to discrete-time and vice versa.

The bilinear transform is a special case of a conformal mapping (namely, the Möbius transformation), often used to convert a transfer function $H_a(s)$ of a linear, time-invariant (LTI) filter in the continuous-time domain (often called an analog filter) to a transfer function $H_d(z)$ of a linear, shift-invariant filter in the discrete-time domain (often called a digital filter although there are analog filters constructed with switched capacitors that are discrete-time filters). It maps positions on the $Re[s] = 0$, in the s-plane to the unit circle, $|z| = 1$, in the z-plane. Other bilinear transforms can be used to warp the frequency response of any discrete-time linear system (for example to approximate the non-linear frequency resolution of the human auditory system) and are implementable in the discrete domain by replacing a system's unit delays (z^{-1}) with first order all-pass filters.

The transform preserves stability and maps every point of the frequency response of the continuous-time filter, $H_a(j\omega_a)$ to a corresponding point in the frequency response of the discrete-time filter, $H_d(e^{j\omega_d T})$ although to a somewhat different frequency, as shown in the Frequency warping section below. This means that for every feature that one sees in the frequency response of the analog filter, there is a corresponding feature, with identical gain and phase shift, in the frequency response of the digital filter but, perhaps, at a somewhat different frequency. This is barely noticeable at low frequencies but is quite evident at frequencies close to the Nyquist frequency.

Discrete-time Approximation

The bilinear transform is a first-order approximation of the natural logarithm function that is an exact mapping of the z-plane to the s-plane. When the Laplace transform is performed on a discrete-time signal (with each element of the discrete-time sequence attached to a correspondingly delayed unit impulse), the result is precisely the Z transform of the discrete-time sequence with the substitution of

$$\begin{aligned} z &= e^{sT} \\ &= \frac{e^{sT/2}}{e^{-sT/2}} \\ &\approx \frac{1+sT/2}{1-sT/2} \end{aligned}$$

where T is the numerical integration step size of the trapezoidal rule used in the bilinear transform derivation; or, in other words, the sampling period. The above bilinear approximation can be solved for s or a similar approximation for $s = (1/T)\ln(z)$ can be performed.

The inverse of this mapping (and its first-order bilinear approximation) is

$$\begin{aligned} s &= \frac{1}{T}\ln(z) \\ &= \frac{2}{T}\left[\frac{z-1}{z+1} + \frac{1}{3}\left(\frac{z-1}{z+1}\right)^3 + \frac{1}{5}\left(\frac{z-1}{z+1}\right)^5 + \frac{1}{7}\left(\frac{z-1}{z+1}\right)^7 + \cdots\right] \\ &\approx \frac{2}{T}\frac{z-1}{z+1} \\ &= \frac{2}{T}\frac{1-z^{-1}}{1+z^{-1}} \end{aligned}$$

The bilinear transform essentially uses this first order approximation and substitutes into the continuous-time transfer function, $H_a(s)$

$$s \leftarrow \frac{2}{T}\frac{z-1}{z+1}.$$

That is

$$H_d(z) = H_a(s)\Big|_{s=\frac{2}{T}\frac{z-1}{z+1}} = H_a\left(\frac{2}{T}\frac{z-1}{z+1}\right)$$

Stability and Minimum-Phase Property Preserved

A continuous-time causal filter is stable if the poles of its transfer function fall in the left half of the complex s-plane. A discrete-time causal filter is stable if the poles of its transfer function fall inside the unit circle in the complex z-plane. The bilinear transform maps the left half of the complex s-plane to the interior of the unit circle in the z-plane. Thus filters designed in the continuous-time domain that are stable are converted to filters in the discrete-time domain that preserve that stability.

Likewise, a continuous-time filter is minimum-phase if the zeros of its transfer function fall in the left half of the complex s-plane. A discrete-time filter is minimum-phase if the zeros of its transfer function fall inside the unit circle in the complex z-plane. Then the same mapping property assures that continuous-time filters that are minimum-phase are converted to discrete-time filters that preserve that property of being minimum-phase.

Example

As an example take a simple low-pass RC filter. This continuous-time filter has a transfer function

$$H_a(s) = \frac{1/sC}{R+1/sC} = \frac{1}{1+RCs}.$$

If we wish to implement this filter as a digital filter, we can apply the bilinear transform by substituting for s the formula above; after some reworking, we get the following filter representation:

$$\begin{aligned} H_d(z) &= H_a\left(\frac{2}{T}\frac{z-1}{z+1}\right) \\ &= \frac{1}{1+RC\left(\frac{2}{T}\frac{z-1}{z+1}\right)} \\ &= \frac{1+z}{(1-2RC/T)+(1+2RC/T)z} \\ &= \frac{1+z^{-1}}{(1+2RC/T)+(1-2RC/T)z^{-1}}. \end{aligned}$$

The coefficients of the denominator are the 'feed-backward' coefficients and the coefficients of the numerator are the 'feed-forward' coefficients used to implement a real-time digital filter.

General second-order Biquad Transformation

It is possible to relate the coefficients of a continuous-time, analog filter with those of a similar discrete-time digital filter created through the bilinear transform process. Transforming a general, second-order continuous-time filter with the given transfer function

$$H_a(s) = \frac{b_0 s^2 + b_1 s + b_2}{a_0 s^2 + a_1 s + a_2} = \frac{b_0 + b_1 s^{-1} + b_2 s^{-2}}{a_0 + a_1 s^{-1} + a_2 s^{-2}}$$

using the bilinear transform (without prewarping any frequency specification) requires the substitution of

$$s \leftarrow K\frac{1-z^{-1}}{1+z^{-1}}$$

where $K \triangleq \frac{2}{T}$.

This results in a discrete-time digital biquad filter with coefficients expressed in terms of the coefficients of the original continuous time filter:

$$H_d(z) = \frac{(b_0K^2 + b_1K + b_2) + (2b_2 - 2b_0K^2)z^{-1} + (b_0K^2 - b_1K + b_2)z^{-2}}{(a_0K^2 + a_1K + a_2) + (2a_2 - 2a_0K^2)z^{-1} + (a_0K^2 - a_1K + a_2)z^{-2}}$$

Normally the constant term in the denominator must be normalized to 1 before deriving the corresponding difference equation. This results in

$$H_d(z) = \frac{\frac{b_0K^2+b_1K+b_2}{a_0K^2+a_1K+a_2} + \frac{2b_2-2b_0K^2}{a_0K^2+a_1K+a_2}z^{-1} + \frac{b_0K^2-b_1K+b_2}{a_0K^2+a_1K+a_2}z^{-2}}{1 + \frac{2a_2-2a_0K^2}{a_0K^2+a_1K+a_2}z^{-1} + \frac{a_0K^2-a_1K+a_2}{a_0K^2+a_1K+a_2}z^{-2}}.$$

The difference equation (using the Direct Form I) is

$$\begin{aligned} y[n] = {} & \frac{b_0K^2+b_1K+b_2}{a_0K^2+a_1K+a_2}\cdot x[n] + \frac{2b_2-2b_0K^2}{a_0K^2+a_1K+a_2}\cdot x[n-1] \\ & + \frac{b_0K^2-b_1K+b_2}{a_0K^2+a_1K+a_2}\cdot x[n-2] - \frac{2a_2-2a_0K^2}{a_0K^2+a_1K+a_2}\cdot y[n-1] \\ & - \frac{a_0K^2-a_1K+a_2}{a_0K^2+a_1K+a_2}\cdot y[n-2]. \end{aligned}$$

Frequency Warping

To determine the frequency response of a continuous-time filter, the transfer function $H_a(s)$ is evaluated at $s = j\omega$ which is on the $j\omega$ axis. Likewise, to determine the frequency response of a discrete-time filter, the transfer function $H_d(z)$ is evaluated at $z = e^{j\omega T}$ which is on the unit circle, $|z| = 1$. When the actual frequency of ω is input to the discrete-time filter designed by use of the bilinear transform, it is desired to know at what frequency, ω_a, for the continuous-time filter that this ω is mapped to.

$$H_d(z) = H_a\left(\frac{2}{T}\frac{z-1}{z+1}\right)$$

$$\begin{aligned} H_d(e^{j\omega T}) &= H_a\left(\frac{2}{T}\frac{e^{j\omega T}-1}{e^{j\omega T}+1}\right) \\ &= H_a\left(\frac{2}{T}\cdot\frac{e^{j\omega T/2}\left(e^{j\omega T/2}-e^{-j\omega T/2}\right)}{e^{j\omega T/2}\left(e^{j\omega T/2}+e^{-j\omega T/2}\right)}\right) \\ &= H_a\left(\frac{2}{T}\cdot\frac{\left(e^{j\omega T/2}-e^{-j\omega T/2}\right)}{\left(e^{j\omega T/2}+e^{-j\omega T/2}\right)}\right) \\ &= H_a\left(j\frac{2}{T}\cdot\frac{\left(e^{j\omega T/2}-e^{-j\omega T/2}\right)/(2j)}{\left(e^{j\omega T/2}+e^{-j\omega T/2}\right)/2}\right) \\ &= H_a\left(j\frac{2}{T}\cdot\frac{\sin(\omega T/2)}{\cos(\omega T/2)}\right) \\ &= H_a\left(j\frac{2}{T}\cdot\tan(\omega T/2)\right) \end{aligned}$$

This shows that every point on the unit circle in the discrete-time filter z-plane, $z = e^{j\omega T}$ is mapped to a point on the $j\omega$ axis on the continuous-time filter s-plane, $s = j\omega_a$. That is, the discrete-time to continuous-time frequency mapping of the bilinear transform is

$$\omega_a = \frac{2}{T}\tan\left(\omega\frac{T}{2}\right)$$

and the inverse mapping is

$$\omega_a = \frac{2}{T}\tan\left(\omega\frac{T}{2}\right)$$

The discrete-time filter behaves at frequency ω_d the same way that the continuous-time filter behaves at frequency $(2/T)\tan(\omega T/2)$. Specifically, the gain and phase shift that the discrete-time filter has at frequency ω is the same gain and phase shift that the continuous-time filter has at frequency $(2/T)\tan(\omega T/2)$. This means that every feature, every "bump" that is visible in the frequency response of the continuous-time filter is also visible in the discrete-time filter, but at a different frequency. For low frequencies (that is, when $\omega \ll 2/T$ or $\omega_a \ll 2/T$), $\omega \approx \omega_a$, then the features are mapped to a slightly different frequency; $\omega_d \approx \omega_a$.

One can see that the entire continuous frequency range

$$-\infty < \omega_a < +\infty$$

is mapped onto the fundamental frequency interval

$$-\frac{\pi}{T} < \omega < +\frac{\pi}{T}.$$

The continuous-time filter frequency $\omega_a = 0$ corresponds to the discrete-time filter frequency $\omega = 0$ and the continuous-time filter frequency $\omega_a = \pm\infty$ correspond to the discrete-time filter frequency $\omega = \pm\pi/T$.

One can also see that there is a nonlinear relationship between ω_a and ω. This effect of the bilinear transform is called frequency warping. The continuous-time filter can be designed to compensate for this frequency warping by setting $\omega_a = \frac{2}{T}\tan\left(\omega\frac{T}{2}\right)$ for every frequency specification that the designer has control over (such as corner frequency or center frequency). This is called pre-warping the filter design.

When designing a digital filter as an approximation of a continuous time filter, the frequency response (both amplitude and phase) of the digital filter can be made to match the frequency response of the continuous filter at frequency ω_0 if the following transform is substituted into the continuous filter transfer function. This is a modified version of Tustin's transform shown above. However, note that this transform becomes the above transform as $\omega_0 \to 0$. That is to say, the above transform causes the digital filter response to match the analog filter response at DC.

$$s \leftarrow \frac{\omega_0}{\tan\frac{\omega_0 T}{2}}\frac{z-1}{z+1}.$$

The main advantage of the warping phenomenon is the absence of aliasing distortion of the frequency response characteristic, such as observed with Impulse invariance. It is necessary, however, to compensate for the frequency warping by pre-warping the given frequency specifications of the continuous-time system. These pre-warped specifications may then be used in the bilinear transform to obtain the desired discrete-time system.

DISCRETE FOURIER TRANSFORM

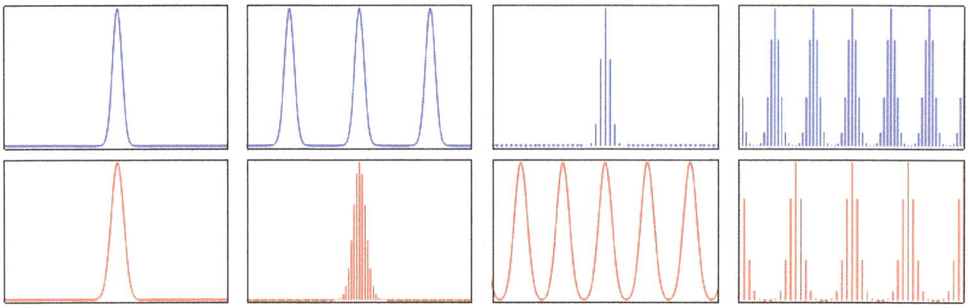

Relationship between the (continuous) Fourier transform and the discrete Fourier transform. Left column: A continuous function (top) and its Fourier transform (bottom). Center-left column: Periodic summation of the original function (top). Fourier transform (bottom) is zero except at discrete points. The inverse transform is a sum of sinusoids called Fourier series. Center-right column: Original function is discretized (multiplied by a Dirac comb) (top). Its Fourier transform (bottom) is a periodic summation (DTFT) of the original transform. Right column: The DFT (bottom) computes discrete samples of the continuous DTFT. The inverse DFT (top) is a periodic summation of the original samples. The FFT algorithm computes one cycle of the DFT and its inverse is one cycle of the DFT inverse.

In mathematics, the discrete Fourier transform (DFT) converts a finite sequence of equally-spaced samples of a function into an equivalent-length sequence of equally-spaced samples of the discrete-time Fourier transform (DTFT), which is a complex-valued function of frequency. The interval at which the DTFT is sampled is the reciprocal of the duration of the input sequence. An inverse DFT is a Fourier series, using the DTFT samples as coefficients of complex sinusoids at the corresponding DTFT frequencies. It has the same sample-values as the original input sequence. The DFT is therefore said to be a frequency domain representation of the original input sequence. If the original sequence spans all the non-zero values of a function, its DTFT is continuous (and periodic), and the DFT provides discrete samples of one cycle. If the original sequence is one cycle of a periodic function, the DFT provides all the non-zero values of one DTFT cycle.

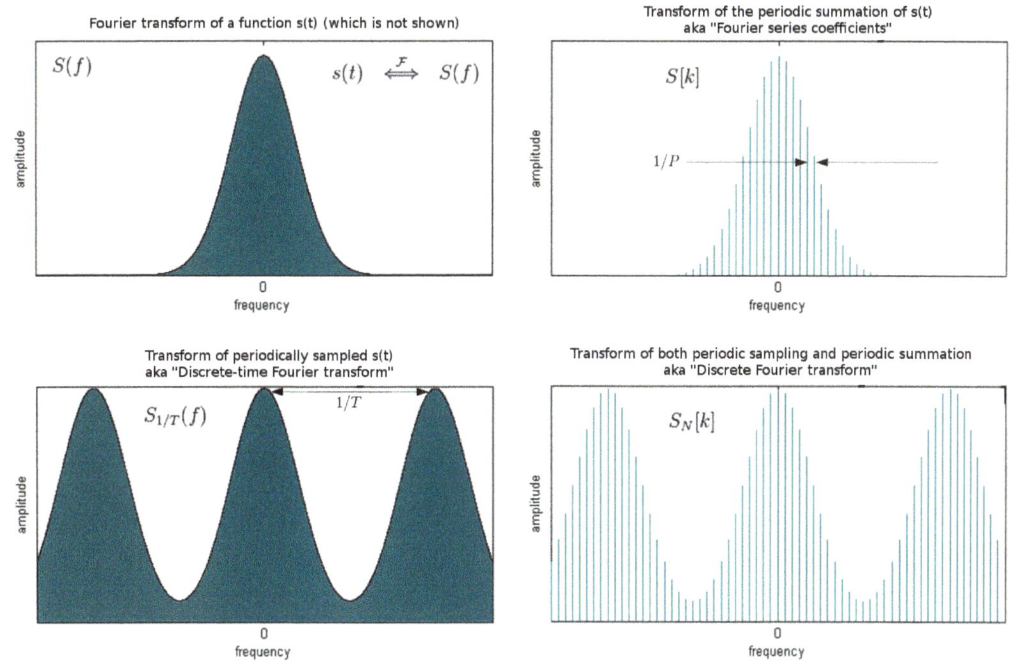

Depiction of a Fourier transform (upper left) and its periodic summation (DTFT) in the lower left corner. The spectral sequences at (a) upper right and (b) lower right are respectively computed from (a) one cycle of the periodic summation of s(t) and (b) one cycle of the periodic summation of the s(nT) sequence. The respective formulas are (a) the Fourier series integral and (b) the **DFT** summation. Its similarities to the original transform, S(f), and its relative computational ease are often the motivation for computing a DFT sequence.

The DFT is the most important discrete transform, used to perform Fourier analysis in many practical applications. In digital signal processing, the function is any quantity or signal that varies over time, such as the pressure of a sound wave, a radio signal, or daily temperature readings, sampled over a finite time interval (often defined by a window function). In image processing, the samples can be the values of pixels along a row or column of a raster image. The DFT is also used to efficiently solve partial differential equations, and to perform other operations such as convolutions or multiplying large integers.

Since it deals with a finite amount of data, it can be implemented in computers by numerical algorithms or even dedicated hardware. These implementations usually employ efficient fast Fourier transform (FFT) algorithms; so much so that the terms "FFT" and "DFT" are often used interchangeably. Prior to its current usage, the "FFT" initialism may have also been used for the ambiguous term "finite Fourier transform.

The sequence of N complex numbers $x_0, x_1, \ldots, x_{N-1}$ is transformed into an N-periodic sequence of complex numbers:

$$X_k \stackrel{\text{def}}{=} \sum_{n=0}^{N-1} x_n \cdot e^{-2\pi i k n/N}, \quad k \in \mathbb{Z} \text{ (integers)}$$

Because of periodicity, the customary domain of k actually computed is [0, N − 1]. That is always the case when the DFT is implemented via the Fast Fourier transform algorithm. But other common domains are [−N/2, N/2 − 1] (N even) and [−(N − 1)/2, (N − 1)/2] (N odd), as when the left and right halves of an FFT output sequence are swapped.

The transform is sometimes denoted by the symbol $\mathcal{F}$, as in $\mathbf{X} = \mathcal{F}\{\mathbf{x}\}$ or $\mathcal{F}(\mathbf{x})$ or $\mathcal{F}\mathbf{x}$.

Eq.1 can be interpreted or derived in various ways, for example:

- It completely describes the discrete-time Fourier transform (DTFT) of an N-periodic sequence, which comprises only discrete frequency components. (Using the DTFT with periodic data)
- It can also provide uniformly spaced samples of the continuous DTFT of a finite length sequence. (Sampling the DTFT)
- It is the cross correlation of the input sequence, x_n, and a complex sinusoid at frequency k/N. Thus it acts like a matched filter for that frequency.
- It is the discrete analogy of the formula for the coefficients of a Fourier series:

$$x_n = \frac{1}{N}\sum_{k=0}^{N-1} X_k \cdot e^{2\pi ikn/N}, \quad n \in \mathbb{Z}$$

which is also N-periodic. In the domain $n \in [0, N-1]$ this is the inverse transform of Eq.1. In this interpretation, each X_k is a complex number that encodes both amplitude and phase of a complex sinusoidal component $(e^{2\pi ikn/N})$ of function x_n.The sinusoid's frequency is k cycles per N samples. Its amplitude and phase are:

$$|X_k|/N = \sqrt{\operatorname{Re}(X_k)^2 + \operatorname{Im}(X_k)^2}/N$$

$$\arg(X_k) = \operatorname{atan2}\big(\operatorname{Im}(X_k), \operatorname{Re}(X_k)\big) = -i\ln\left(\frac{X_k}{|X_k|}\right),$$

where atan2 is the two-argument form of the arctan function.

The normalization factor multiplying the DFT and IDFT (here 1 and 1/N) and the signs of the exponents are merely conventions, and differ in some treatments. The only requirements of these conventions are that the DFT and IDFT have opposite-sign exponents and that the product of their normalization factors be 1/N. A normalization of for both the DFT and IDFT, for instance, makes the transforms unitary.

In the following discussion the terms sequence and vector will be considered interchangeable.

Using Euler's formula, the DFT formulae can be converted to the trigonometric forms sometimes used in engineering and computer science:

Fourier Transform:

$$X_k = \sum_{n=0}^{N-1} x_n \cdot (\cos(-2\pi k \frac{n}{N}) + i \sin(-2\pi k \frac{n}{N})), \quad k \in \mathbb{Z}$$

Inverse Fourier Transform:

$$x_n = \frac{1}{N} \sum_{k=0}^{N-1} X_k \cdot (\cos(2\pi k \frac{n}{N}) + i \sin(2\pi k \frac{n}{N})), \quad n \in \mathbb{Z}$$

N = number of time samples we have

n = current sample we're considering (0, ..., N − 1)

xn = value of the signal at time n

k = current frequency we're considering (0 Hertz up to N-1 Hertz)

Xk = amount of frequency k in the signal (Amplitude and Phase, a complex number)

Properties

Completeness

The discrete Fourier transform is an invertible, linear transformation

$\mathcal{F}: \mathbb{C}^N \to \mathbb{C}^N$

with C denoting the set of complex numbers. In other words, for any N > 0, an N-dimensional complex vector has a DFT and an IDFT which are in turn N-dimensional complex vectors.

Orthogonality

The vectors $u_k = \left[e^{\frac{2\pi i}{N} kn} \mid n = 0, 1, \ldots, N-1 \right]^T$ form an orthogonal basis over the set of N-dimensional complex vectors:

$$u_k^T u_{k'}^* = \sum_{n=0}^{N-1} \left(e^{\frac{2\pi i}{N} kn} \right) \left(e^{\frac{2\pi i}{N} (-k')n} \right) = \sum_{n=0}^{N-1} e^{\frac{2\pi i}{N} (k-k')n} = N\, \delta_{kk'}$$

where $\delta_{kk'}$ is the Kronecker delta. (In the last step, the summation is trivial if $k = k'$, where it is 1+1+⋯=N, and otherwise is a geometric series that can be explicitly summed to obtain zero.) This orthogonality condition can be used to derive the formula for the IDFT from the definition of the DFT, and is equivalent to the unitarity property below.

The Plancherel theorem and Parseval's theorem

If Xk and Yk are the DFTs of xn and yn respectively then the Parseval's theorem states:

$$\sum_{n=0}^{N-1} x_n y_n^* = \frac{1}{N}\sum_{k=0}^{N-1} X_k Y_k^*$$

where the star denotes complex conjugation. Plancherel theorem is a special case of the Parseval's theorem and states:

$$\sum_{n=0}^{N-1} |x_n|^2 = \frac{1}{N}\sum_{k=0}^{N-1} |X_k|^2.$$

These theorems are also equivalent to the unitary condition below.

Periodicity

The periodicity can be shown directly from the definition:

$$X_{k+N} \stackrel{\text{def}}{=} \sum_{n=0}^{N-1} x_n e^{-\frac{2\pi i}{N}(k+N)n} = \sum_{n=0}^{N-1} x_n e^{-\frac{2\pi i}{N}kn} \underbrace{e^{-2\pi i n}}_{1} = \sum_{n=0}^{N-1} x_n e^{-\frac{2\pi i}{N}kn} = X_k.$$

Similarly, it can be shown that the IDFT formula leads to a periodic extension.

Shift Theorem

Multiplying x_n by a *linear phase* $e^{\frac{2\pi i}{N}nm}$ for some integer m corresponds to a circular shift of the output X_k: X_k is replaced by X_{k-m}, where the subscript is interpreted modulo N (i.e., periodically). Similarly, a circular shift of the input x_n corresponds to multiplying the output X_k by a linear phase. Mathematically, if $\{x_n\}$ represents the vector x then

if $\mathcal{F}(\{x_n\})_k = X_k$

then

$$\mathcal{F}^{-1}\{\mathbf{X}\cdot\mathbf{Y}\}_n = \sum_{l=0}^{N-1} x_l \cdot (y_N)_{n-l} \stackrel{\text{def}}{=} (\mathbf{x} * \mathbf{y_N})_n\ ,$$

and

$$(\mathbf{y_N})_n \stackrel{\text{def}}{=} \sum_{p=-\infty}^{\infty} y_{(n-pN)} = y_{n(modN)}.$$

Circular Convolution Theorem and Cross-Correlation Theorem

The convolution theorem for the discrete-time Fourier transform indicates that a convolution of two infinite sequences can be obtained as the inverse transform of the product of the individual transforms. An important simplification occurs when the sequences are of finite length, N. In terms of the DFT and inverse DFT, it can be written as follows:

$$\mathcal{F}^{-1}\{\mathbf{X}\cdot\mathbf{Y}\}_n = \sum_{l=0}^{N-1} x_l \cdot (y_N)_{n-l} \stackrel{\text{def}}{=} (\mathbf{x} * \mathbf{y_N})_n ,$$

which is the convolution of the x sequence with a y sequence extended by periodic summation:

$$(\mathbf{y_N})_n \stackrel{\text{def}}{=} \sum_{p=-\infty}^{\infty} y_{(n-pN)} = y_{n(modN)}.$$

Similarly, the cross-correlation of $\mathbf{x}$ and $\mathbf{y_N}$ is given by:

$$\mathcal{F}^{-1}\{\mathbf{X}^*\cdot\mathbf{Y}\}_n = \sum_{l=0}^{N-1} x_l^* \cdot (y_N)_{n+l} \stackrel{\text{def}}{=} (\mathbf{x} \star \mathbf{y_N})_n .$$

When either sequence contains a string of zeros, of length L, L+1 of the circular convolution outputs are equivalent to values of $\mathbf{x}*\mathbf{y}$ Methods have also been developed to use this property as part of an efficient process that constructs $\mathbf{x}*\mathbf{y}$ with an $\mathbf{x}$ or $\mathbf{y}$ sequence potentially much longer than the practical transform size (N). Two such methods are called overlap-save and overlap-add. The efficiency results from the fact that a direct evaluation of either summation (above) requires $O(N^2)$ operations for an output sequence of length N. An indirect method, using transforms, can take advantage of the $O(N \log N)$ efficiency of the fast Fourier transform (FFT) to achieve much better performance. Furthermore, convolutions can be used to efficiently compute DFTs via Rader's FFT algorithm and Bluestein's FFT algorithm.

Convolution theorem duality

It can also be shown that:

$$\mathcal{F}\{\mathbf{x}\cdot\mathbf{y}\}_k \stackrel{\text{def}}{=} \sum_{n=0}^{N-1} x_n \cdot y_n \cdot e^{-\frac{2\pi i}{N}kn}$$

$$= \frac{1}{N}(\mathbf{X} * \mathbf{Y_N})_k,$$

hich is the circular convolution of X and Y.

Trigonometric Interpolation Polynomial

The trigonometric interpolation polynomial

$$p(t) = \frac{1}{N}\left[X_0 + X_1 e^{2\pi i t} + \cdots + X_{N/2-1} e^{(N/2-1)2\pi i t} + X_{N/2}\cos(N\pi t) + X_{N/2+1} e^{(-N/2+1)2\pi i t} + \cdots + X_{N-1} e^{-2\pi i t}\right]$$

for N even ,

$$p(t) = \frac{1}{N}\left[X_0 + X_1 e^{2\pi i t} + \cdots + X_{\lfloor N/2\rfloor} e^{\lfloor N/2\rfloor 2\pi i t} + X_{\lfloor N/2\rfloor+1} e^{-\lfloor N/2\rfloor 2\pi i t} + \cdots + X_{N-1} e^{-2\pi i t}\right]$$

for N odd,

where the coefficients Xk are given by the DFT of xn above, satisfies the interpolation property $p(n/N) = x_n$ for $n = 0, \ldots, N-1$.

For even N, notice that the Nyquist component $\frac{X_{N/2}}{N}\cos(N\pi t)$ is handled specially.

This interpolation is not unique: aliasing implies that one could add N to any of the complex-sinusoid frequencies (e.g. changing e^{-it} to $e^{i(N-1)t}$) without changing the interpolation property, but giving different values in between the x_n points. The choice above, however, is typical because it has two useful properties. First, it consists of sinusoids whose frequencies have the smallest possible magnitudes: the interpolation is bandlimited. Second, if the x_n are real numbers, then $p(t)$ is real as well.

In contrast, the most obvious trigonometric interpolation polynomial is the one in which the frequencies range from 0 to $N-1$ (instead of roughly $-N/2$ to $+N/2$ as above), similar to the inverse DFT formula. This interpolation does not minimize the slope, and is not generally real-valued for real its use is a common mistake.

The unitary DFT

Another way of looking at the DFT is to note that in the above discussion, the DFT can be expressed as the DFT matrix, a Vandermonde matrix, introduced by Sylvester in 1867,

$$\mathbf{F} = \begin{bmatrix} \omega_N^{0\cdot 0} & \omega_N^{0\cdot 1} & \cdots & \omega_N^{0\cdot(N-1)} \\ \omega_N^{1\cdot 0} & \omega_N^{1\cdot 1} & \cdots & \omega_N^{1\cdot(N-1)} \\ \vdots & \vdots & \ddots & \vdots \\ \omega_N^{(N-1)\cdot 0} & \omega_N^{(N-1)\cdot 1} & \cdots & \omega_N^{(N-1)\cdot(N-1)} \end{bmatrix}$$

where

$$\omega_N = e^{-2\pi i/N}$$

is a primitive Nth root of unity.

The inverse transform is then given by the inverse of the above matrix,

$$\mathbf{F}^{-1} = \frac{1}{N}\mathbf{F}^*$$

With unitary normalization constants $1/\sqrt{N}$, the DFT becomes a unitary transformation, defined by a unitary matrix:

$$\mathbf{U} = \mathbf{F}/\sqrt{N}$$
$$\mathbf{U}^{-1} = \mathbf{U}^*$$
$$|\det(\mathbf{U})| = 1$$

where det() is the determinant function. The determinant is the product of the eigenvalues, which are always ± 1 or $\pm i$ as described below. In a real vector space, a unitary transformation can be thought of as simply a rigid rotation of the coordinate system, and all of the properties of a rigid rotation can be found in the unitary DFT.

The orthogonality of the DFT is now expressed as an orthonormality condition (which arises in many areas of mathematics as described in root of unity):

$$\sum_{m=0}^{N-1} U_{km} U_{mn}^{*} = \delta_{kn}$$

If X is defined as the unitary DFT of the vector x, then

$$X_k = \sum_{n=0}^{N-1} U_{kn} x_n$$

and the Plancherel theorem is expressed as

$$\sum_{n=0}^{N-1} |x_n|^2 = \sum_{k=0}^{N-1} |X_k|^2$$

If we view the DFT as just a coordinate transformation which simply specifies the components of a vector in a new coordinate system, then the above is just the statement that the dot product of two vectors is preserved under a unitary DFT transformation. For the special case $\mathbf{x} = \mathbf{y}$, this implies that the length of a vector is preserved as well—this is just Parseval's theorem,

$$\mathcal{F}^{-1}(\{x_n\}) = \mathcal{F}(\{x_{N-n}\})/N$$

A consequence of the circular convolution theorem is that the DFT matrix F diagonalizes any circulant matrix.

Expressing the inverse DFT in terms of the DFT

A useful property of the DFT is that the inverse DFT can be easily expressed in terms of the (forward) DFT, via several well-known "tricks". (For example, in computations, it is often convenient to only implement a fast Fourier transform corresponding to one transform direction and then to get the other transform direction from the first.)

First, we can compute the inverse DFT by reversing all but one of the inputs (Duhamel et al., 1988):

$$\mathcal{F}^{-1}(\{x_n\}) = \mathcal{F}(\{x_{N-n}\})/N$$

(As usual, the subscripts are interpreted modulo N; thus, for $n = 0$, we have $x_{N-0} = x_0$.)

Second, one can also conjugate the inputs and outputs:

$$\mathcal{F}^{-1}(\mathbf{x}) = \mathcal{F}(\mathbf{x}^*)^*/N$$

Third, a variant of this conjugation trick, which is sometimes preferable because it requires no modification of the data values, involves swapping real and imaginary parts (which can be done on a computer simply by modifying pointers). Define swap (x_n) as x_n with its real and imaginary parts swapped—that is, if $x_n = a + bi$ then swap(x_n) is $b + ai$. Equivalently, swap(x_n) then swap ix_n^*. Then

$$\mathcal{F}^{-1}(\mathbf{x}) = \mathrm{swap}(\mathcal{F}(\mathrm{swap}(\mathbf{x})))/N$$

That is, the inverse transform is the same as the forward transform with the real and imaginary parts swapped for both input and output, up to a normalization (Duhamel et al., 1988).

The conjugation trick can also be used to define a new transform, closely related to the DFT, that is involutory—that is, which is its own inverse. In particular, $T(\mathbf{x}) = \mathcal{F}(\mathbf{x}^*)/\sqrt{N}$ is clearly its own inverse: $T(T(\mathbf{x})) = \mathbf{x}$. A closely related involutory transformation (by a factor of (1+i) /√2) is $H(\mathbf{x}) = \mathcal{F}((1+i)\mathbf{x}^*)/\sqrt{2N}$, since the $(1+i)$ factors in $H(H(\mathbf{x}))$ cancel the 2. For real inputs x, the real part of $H(\mathbf{x})$is none other than the discrete Hartley transform, which is also involutory.

Eigenvalues and Eigenvectors

The eigenvalues of the DFT matrix are simple and well-known, whereas the eigenvectors are complicated, not unique, and are the subject of ongoing research.

Consider the unitary form {\displaystyle \mathbf **U** defined above for the DFT of length N, where

$$\mathbf{U}_{m,n} = \frac{1}{\sqrt{N}}\omega_N^{(m-1)(n-1)} = \frac{1}{\sqrt{N}}e^{-\frac{2\pi i}{N}(m-1)(n-1)}.$$

This matrix satisfies the matrix polynomial equation:

$$\mathbf{U}^4 = \mathbf{I}.$$

This can be seen from the inverse properties above: operating U twice gives the original data in reverse order, so operating U four times gives back the original data and is thus the identity matrix. This means that the eigenvalues U satisfy the equation:

$$\lambda^4 = 1.$$

Therefore, the eigenvalues of U are the fourth roots of unity: λ is +1, −1, +*i*, or −*i*.

Since there are only four distinct eigenvalues for this N matrix, they have some multiplicity. The multiplicity gives the number of linearly independent eigenvectors corresponding to each eigenvalue. (Note that there are N independent eigenvectors; a unitary matrix is never defective.)

The problem of their multiplicity was solved by McClellan and Parks (1972), although it was later shown to have been equivalent to a problem solved by Gauss (Dickinson and Steiglitz, 1982). The multiplicity depends on the value of N modulo 4, and is given by the following table:

size N	$\lambda = +1$	$\lambda = -1$	$\lambda = -i$	$\lambda = +i$
$4m$	$m + 1$	m	m	$m - 1$
$4m + 1$	$m + 1$	m	m	m
$4m + 2$	$m + 1$	$m + 1$	m	m
$4m + 3$	$m + 1$	$m + 1$	$m + 1$	m

Multiplicities of the eigenvalues λ of the unitary **DFT** matrix **U** as a function of the transform size N (in terms of an integer m).

size N	$\lambda = +1$	$\lambda = -1$	$\lambda = -i$	$\lambda = +i$
4m	m + 1	m	m	m − 1
4m + 1	m + 1	m	m	m
4m + 2	m + 1	m + 1	m	m
4m + 3	m + 1	m + 1	m + 1	m

Multiplicities of the eigenvalues λ of the unitary DFT matrix U as a function of the transform size *N* (in terms of an integer *m*).

Otherwise stated, the characteristic polynomial of U is:

$$\det(\lambda I - \mathbf{U}) = (\lambda - 1)^{\left\lfloor \frac{N+4}{4} \right\rfloor} (\lambda + 1)^{\left\lfloor \frac{N+2}{4} \right\rfloor} (\lambda + i)^{\left\lfloor \frac{N+1}{4} \right\rfloor} (\lambda - i)^{\left\lfloor \frac{N-1}{4} \right\rfloor}.$$

No simple analytical formula for general eigenvectors is known. Moreover, the eigenvectors are not unique because any linear combination of eigenvectors for the same eigenvalue is also an eigenvector for that eigenvalue. Various researchers have proposed different choices of eigenvectors, selected to satisfy useful properties like orthogonality and to have "simple" forms (e.g., McClellan and Parks, 1972; Dickinson and Steiglitz, 1982; Grünbaum, 1982; Atakishiyev and Wolf, 1997; Candan et al., 2000; Hanna et al., 2004; Gurevich and Hadani, 2008).

A straightforward approach is to discretize an eigenfunction of the continuous Fourier transform, of which the most famous is the Gaussian function. Since periodic summation of the function means discretizing its frequency spectrum and discretization means periodic summation of the spectrum, the discretized and periodically summed Gaussian function yields an eigenvector of the discrete transform:

- $$F(m) = \sum_{k\in\mathbb{Z}} \exp\left(-\frac{\pi\cdot(m+N\cdot k)^2}{N}\right)$$

A closed form expression for the series is not known, but it converges rapidly.

Two other simple closed-form analytical eigenvectors for special DFT period N were found (Kong, 2008):

For DFT period N = 2L + 1 = 4K +1, where K is an integer, the following is an eigenvector of DFT:

- $$F(m) = \prod_{s=K+1}^{L}\left[\cos\left(\frac{2\pi}{N}m\right) - \cos\left(\frac{2\pi}{N}s\right)\right]$$

For DFT period N = 2L = 4K, where K is an integer, the following is an eigenvector of DFT:

- $$\sum_{n=0}^{N-1} |X_n|^2 = 1 \,,$$

The choice of eigenvectors of the DFT matrix has become important in recent years in order to define a discrete analogue of the fractional Fourier transform—the DFT matrix can be taken to fractional powers by exponentiating the eigenvalues (e.g., Rubio and Santhanam, 2005). For the continuous Fourier transform, the natural orthogonal eigenfunctions are the Hermite functions, so various discrete analogues of these have been employed as the eigenvectors of the DFT, such as the Kravchuk polynomials (Atakishiyev and Wolf, 1997). The "best" choice of eigenvectors to define a fractional discrete Fourier transform remains an open question, however.

Uncertainty principle

If the random variable Xk is constrained by

$$\sum_{n=0}^{N-1} |X_n|^2 = 1 \,,$$

then

$$P_n = |X_n|^2$$

may be considered to represent a discrete probability mass function of n, with an associated probability mass function constructed from the transformed variable,

$$Q_m = N|x_m|^2$$

For the case of continuous functions $P(x)$ and $Q(k)$, the Heisenberg uncertainty principle states that

$$D_0(X)D_0(x) \geq \frac{1}{16\pi^2}$$

where $D_0(X)$ and $D_0(x)$ are the variances of $|X|^2$ and $|x|^2$ respectively, with the equality attained in the case of a suitably normalized Gaussian distribution. Although the variances may be analogously defined for the DFT, an analogous uncertainty principle is not useful, because the uncertainty will not be shift-invariant. Still, a meaningful uncertainty principle has been introduced by Massar and Spindel.

However, the Hirschman entropic uncertainty will have a useful analog for the case of the DFT. The Hirschman uncertainty principle is expressed in terms of the Shannon entropy of the two probability functions.

In the discrete case, the Shannon entropies are defined as

$$H(X) = -\sum_{n=0}^{N-1} P_n \ln P_n$$

and

$$H(x) = -\sum_{m=0}^{N-1} Q_m \ln Q_m ,$$

and the entropic uncertainty principle becomes

$$H(X) + H(x) \geq \ln(N) .$$

The equality is obtained for P_n equal to translations and modulations of a suitably normalized Kronecker comb of period A where A is any exact integer divisor of N. The probability mass function Q_m will then be proportional to a suitably translated Kronecker comb of period $B = N/A$.

There is also a well-known deterministic uncertainty principle that uses signal sparsity (or the number of non-zero coefficients). Let $\|x\|_0$ and $\|X\|_0$ be the number of non-zero elements of the time and frequency sequences $x_0, x_1, \ldots, x_{N-1}$ and $X_0, X_1, \ldots, X_{N-1}$, respectively. Then,

$$N \leq \|x\|_0 \cdot \|X\|_0 .$$

As an immediate consequence of the inequality of arithmetic and geometric means, one also has $2\sqrt{N} \leq \|x\|_0 + \|X\|_0$. Both uncertainty principles were shown to be tight for specifically-chosen "picket-fence" sequences (discrete impulse trains), and find practical use for signal recovery applications.

The real-Input DFT

If $x_0, \ldots, x_{N-1}$ are real numbers, as they often are in practical applications, then the DFT obeys the symmetry:

$X_{N-k} \equiv X_{-k} = X_k^*$, where X^* denotes complex conjugation.

It follows that X0 and XN/2 are real-valued, and the remainder of the DFT is completely specified by just N/2-1 complex numbers.

Generalized DFT (Shifted and Non-Linear Phase)

It is possible to shift the transform sampling in time and/or frequency domain by some real shifts a and b, respectively. This is sometimes known as a generalized DFT (or GDFT), also called the shifted DFT or offset DFT, and has analogous properties to the ordinary DFT:

$$X_k = \sum_{n=0}^{N-1} x_n e^{-\frac{2\pi i}{N}(k+b)(n+a)} \qquad k = 0, \ldots, N-1.$$

Most often, shifts of 1/2 (half a sample) are used. While the ordinary DFT corresponds to a periodic signal in both time and frequency domains, a=1/2 produces a signal that is anti-periodic in frequency domain $(X_{k+N} = -X_k)$ and vice versa for $b = 1/2$. Thus, the specific case of {\displaystyle a=b=1/2} is known as an odd-time odd-frequency discrete Fourier transform (or O2DFT). Such shifted transforms are most often used for symmetric data, to represent different boundary symmetries, and for real-symmetric data they correspond to different forms of the discrete cosine and sine transforms.

Another interesting choice is $a = b = -(N-1)/2$, which is called the centered DFT (or CDFT). The centered DFT has the useful property that, when N is a multiple of four, all four of its eigenvalues (see above) have equal multiplicities (Rubio and Santhanam, 2005)

The term GDFT is also used for the non-linear phase extensions of DFT. Hence, GDFT method provides a generalization for constant amplitude orthogonal block transforms including linear and non-linear phase types. GDFT is a framework to improve time and frequency domain properties of the traditional DFT, e.g. auto/cross-correlations, by the addition of the properly designed phase shaping function (non-linear, in general) to the original linear phase functions (Akansu and Agirman-Tosun, 2010).

The discrete Fourier transform can be viewed as a special case of the z-transform, evaluated on the unit circle in the complex plane; more general z-transforms correspond to complex shifts a and b above.

Multidimensional DFT

The ordinary DFT transforms a one-dimensional sequence or array x_n that is a function of exactly one discrete variable n. The multidimensional DFT of a multidimensional array $x_{n_1,n_2,\ldots,n_d}$ that is a function of d discrete variables $n_\ell = 0, 1, \ldots, N_\ell - 1 \text{ for } \ell \text{ in } 1, 2, \ldots, d$ is defined by:

$$X_{k_1,k_2,\ldots,k_d} = \sum_{n_1=0}^{N_1-1} \left(\omega_{N_1}^{k_1 n_1} \sum_{n_2=0}^{N_2-1} \left(\omega_{N_2}^{k_2 n_2} \cdots \sum_{n_d=0}^{N_d-1} \omega_{N_d}^{k_d n_d} \cdot x_{n_1,n_2,\ldots,n_d} \right) \right)$$

where $\omega_{N_\ell} = \exp(-2\pi i/N_\ell)$ as above and the d output indices run from $k_\ell = 0, 1, \ldots, N_\ell - 1$. This is more compactly expressed in vector notation, where we define

$\mathbf{n} = (n_1, n_2, \ldots, n_d)$ and $\mathbf{k} = (k_1, k_2, \ldots, k_d)$ and as d-dimensional vectors of indices from 0 to $\mathbf{N} - 1$, which we define as $\mathbf{N} - 1 = (N_1 - 1, N_2 - 1, \ldots, N_d - 1)$:

$$X_{\mathbf{k}} = \sum_{\mathbf{n}=0}^{\mathbf{N}-1} e^{-2\pi i \mathbf{k}\cdot(\mathbf{n}/\mathbf{N})} x_{\mathbf{n}} ,$$

where the division $\mathbf{n}/\mathbf{N}$ is defined as $\mathbf{n}/\mathbf{N} = (n_1/N_1, \ldots, n_d/N_d)$ to be performed element-wise, and the sum denotes the set of nested summations above.

The inverse of the multi-dimensional DFT is, analogous to the one-dimensional case, given by:

$$x_{\mathbf{n}} = \frac{1}{\prod_{\ell=1}^{d} N_\ell} \sum_{\mathbf{k}=0}^{\mathbf{N}-1} e^{2\pi i \mathbf{n}\cdot(\mathbf{k}/\mathbf{N})} X_{\mathbf{k}}$$

As the one-dimensional DFT expresses the input x_n as a superposition of sinusoids, the multidimensional DFT expresses the input as a superposition of plane waves, or multidimensional sinusoids. The direction of oscillation in space is $\mathbf{k}/\mathbf{N}$. The amplitudes are $X_{\mathbf{k}}$. This decomposition is of great importance for everything from digital image processing (two-dimensional) to solving partial differential equations. The solution is broken up into plane waves.

The multidimensional DFT can be computed by the composition of a sequence of one-dimensional DFTs along each dimension. In the two-dimensional case x_{n_1,n_2} the N_1 independent DFTs of the rows (i.e., along n_2 are computed first to form a new array y_{n_1,k_2}. Then the N_2 independent DFTs of y along the columns (along n_1) are computed to form the final result X_{k_1,k_2}. Alternatively the columns can be computed first and then the rows. The order is immaterial because the nested summations above commute.

An algorithm to compute a one-dimensional DFT is thus sufficient to efficiently compute a multidimensional DFT. This approach is known as the row-column algorithm. There are also intrinsically multidimensional FFT algorithms.

The real-input Multidimensional DFT

For input data $x_{n_1,n_2,\ldots,n_d}$ consisting of real numbers, the DFT outputs have a conjugate symmetry similar to the one-dimensional case above:

$$X_{k_1,k_2,\ldots,k_d} = X^*_{N_1-k_1,N_2-k_2,\ldots,N_d-k_d},$$

where the star again denotes complex conjugation and the ℓ-th subscript is again interpreted modulo N_ℓ (for $\ell = 1, 2, \ldots, d$)..

Applications

The DFT has seen wide usage across a large number of fields; we only sketch a few examples below (see also the references at the end). All applications of the DFT depend crucially on the availability of a fast algorithm to compute discrete Fourier transforms and their inverses, a fast Fourier transform.

Spectral Analysis

When the DFT is used for signal spectral analysis, the $\{x_n\}$ sequence usually represents a finite set of uniformly spaced time-samples of some signal $x(t)$, where t represents time. The conversion from continuous time to samples (discrete-time) changes the underlying Fourier transform of x(t) into a discrete-time Fourier transform (DTFT), which generally entails a type of distortion called aliasing. Choice of an appropriate sample-rate (see Nyquist rate) is the key to minimizing that distortion. Similarly, the conversion from a very long (or infinite) sequence to a manageable size entails a type of distortion called leakage, which is manifested as a loss of detail (a.k.a. resolution) in the DTFT. Choice of an appropriate sub-sequence length is the primary key to minimizing that effect. When the available data (and time to process it) is more than the amount needed to attain the desired frequency resolution, a standard technique is to perform multiple DFTs, for example to create a spectrogram. If the desired result is a power spectrum and noise or randomness is present in the data, averaging the magnitude components of the multiple DFTs is a useful procedure to reduce the variance of the spectrum (also called a periodogram in this context); two examples of such techniques are the Welch method and the Bartlett method; the general subject of estimating the power spectrum of a noisy signal is called spectral estimation.

A final source of distortion (or perhaps illusion) is the DFT itself, because it is just a discrete sampling of the DTFT, which is a function of a continuous frequency domain. That can be mitigated by increasing the resolution of the DFT. That procedure is illustrated at Sampling the DTFT.

- The procedure is sometimes referred to as zero-padding, which is a particular implementation used in conjunction with the fast Fourier transform (FFT) algorithm. The inefficiency of performing multiplications and additions with zero-valued "samples" is more than offset by the inherent efficiency of the FFT.
- As already noted, leakage imposes a limit on the inherent resolution of the DTFT. So there is a practical limit to the benefit that can be obtained from a fine-grained DFT.

Filter Bank

Data Compression

The field of digital signal processing relies heavily on operations in the frequency domain (i.e. on the Fourier transform). For example, several lossy image and sound compression methods employ the discrete Fourier transform: the signal is cut into short segments, each is transformed, and then the Fourier coefficients of high frequencies, which are assumed to be unnoticeable, are discarded. The decompressor computes the inverse transform based on this reduced number of Fourier coefficients. (Compression applications often use a specialized form of the DFT, the discrete cosine transform or sometimes the modified discrete cosine transform.) Some relatively recent compression algorithms, however, use wavelet transforms, which give a more uniform compromise between time and frequency domain than obtained by chopping data into segments and transforming each segment. In the case of JPEG2000, this avoids the spurious image features that appear when images are highly compressed with the original JPEG.

Partial Differential Equations

Discrete Fourier transforms are often used to solve partial differential equations, where again the DFT is used as an approximation for the Fourier series (which is recovered in the limit of infinite N). The advantage of this approach is that it expands the signal in complex exponentials e^{inx}, which are eigenfunctions of differentiation: $\mathrm{d}(e^{inx})/\mathrm{d}x = ine^{inx}$. Thus, in the Fourier representation, differentiation is simple—we just multiply by in. (Note, however, that the choice of n is not unique due to aliasing; for the method to be convergent, a choice similar to that in the trigonometric interpolation section above should be used.) A linear differential equation with constant coefficients is transformed into an easily solvable algebraic equation. One then uses the inverse DFT to transform the result back into the ordinary spatial representation. Such an approach is called a spectral method.

Polynomial Multiplication

Suppose we wish to compute the polynomial product c(x) = a(x) · b(x). The ordinary product expression for the coefficients of c involves a linear (acyclic) convolution, where indices do not "wrap around." This can be rewritten as a cyclic convolution by taking the coefficient vectors for a(x) and b(x) with constant term first, then appending zeros so that the resultant coefficient vectors a and b have dimension d > deg(a(x)) + deg(b(x)). Then,

$$\mathbf{c} = \mathbf{a} * \mathbf{b}$$

Where c is the vector of coefficients for c(x), and the convolution operator * is defined so

$$c_n = \sum_{m=0}^{d-1} a_m b_{n-m \bmod d} \qquad n = 0, 1 \ldots, d-1$$

But convolution becomes multiplication under the DFT:

$$\mathcal{F}(\mathbf{c}) = \mathcal{F}(\mathbf{a})\mathcal{F}(\mathbf{b})$$

Here the vector product is taken elementwise. Thus the coefficients of the product polynomial c(x) are just the terms 0, ..., deg(a(x)) + deg(b(x)) of the coefficient vector

$$\mathbf{c} = \mathcal{F}^{-1}(\mathcal{F}(\mathbf{a})\mathcal{F}(\mathbf{b}))$$

With a fast Fourier transform, the resulting algorithm takes O (N log N) arithmetic operations. Due to its simplicity and speed, the Cooley–Tukey FFT algorithm, which is limited to composite sizes, is often chosen for the transform operation. In this case, d should be chosen as the smallest integer greater than the sum of the input polynomial degrees that is factorizable into small prime factors (e.g. 2, 3, and 5, depending upon the FFT implementation).

Multiplication of Large Integers

The fastest known algorithms for the multiplication of very large integers use the polynomial multiplication method outlined above. Integers can be treated as the value of a polynomial evaluated specifically at the number base, with the coefficients of the polynomial corresponding to the digits in that base. After polynomial multiplication, a relatively low-complexity carry-propagation step completes the multiplication.

Convolution

When data is convolved with a function with wide support, such as for downsampling by a large sampling ratio, because of the Convolution theorem and the FFT algorithm, it may be faster to transform it, multiply pointwise by the transform of the filter and then reverse transform it. Alternatively, a good filter is obtained by simply truncating the transformed data and re-transforming the shortened data set.

Some discrete Fourier Transform Pairs

Some DFT pairs

$x_n = \frac{1}{N}\sum_{k=0}^{N-1} X_k e^{i2\pi kn/N}$	$X_k = \sum_{n=0}^{N-1} x_n e^{-i2\pi kn/N}$	Note
$x_n e^{i2\pi n\ell/N}$	$X_{k-\ell}$	Shift theorem
$x_{n-\ell}$	$X_k e^{-i2\pi k\ell/N}$	
$x_n \in \mathbb{R}$	$X_k = X^*_{N-k}$	Real DFT
a^n	$\begin{cases} N & \text{if } a = e^{i2\pi k/N} \\ \frac{1-a^N}{1-a\,e^{-i2\pi k/N}} & \text{otherwise} \end{cases}$	from the geometric progression formula
$\binom{N-1}{n}$	$\left(1+e^{-i2\pi k/N}\right)^{N-1}$	from the binomial theorem
$\begin{cases} \frac{1}{W} & \text{if } 2n < W \text{ or } 2(N-n) < W \\ 0 & \text{otherwise} \end{cases}$	$\begin{cases} 1 & \text{if } k = 0 \\ \frac{\sin\left(\frac{\pi Wk}{N}\right)}{W\sin\left(\frac{\pi k}{N}\right)} & \text{otherwise} \end{cases}$	x_n is a rectangular window function of W points centered on n=0, where W is an odd integer, and X_k is a sinc-like function (specifically, X_k is a Dirichlet kernel)
$\sum_{j\in\mathbb{Z}} \exp\left(-\frac{\pi}{cN}\cdot(n+N\cdot j)^2\right)$	$\sqrt{cN}\cdot\sum_{j\in\mathbb{Z}} \exp\left(-\frac{\pi c}{N}\cdot(k+N\cdot j)^2\right)$	Discretization and periodic summation of the scaled Gaussian functions for $c > 0$. Since either c or $\frac{1}{c}$ is larger than one and thus warrants fast convergence of one of the two series, for large c you may choose to compute the frequency spectrum and convert to the time domain using the discrete Fourier transform.

Generalizations

Representation Theory

The DFT can be interpreted as the complex-valued representation theory of the finite cyclic group. In other words, a sequence of n complex numbers can be thought of as an element of n-dimensional complex space Cn or equivalently a function f from the finite cyclic group of order n to the complex numbers, Zn → C. So f is a class function on the finite cyclic group, and thus can be expressed as a linear combination of the irreducible characters of this group, which are the roots of unity.

From this point of view, one may generalize the DFT to representation theory generally, or more narrowly to the representation theory of finite groups.

More narrowly still, one may generalize the DFT by either changing the target (taking values in a field other than the complex numbers), or the domain (a group other than a finite cyclic group), as detailed in the sequel.

Other Fields

Many of the properties of the DFT only depend on the fact that $e^{-\frac{2\pi i}{N}}$ is a primitive root of unity, sometimes denoted W_N (so that $\omega_N^N = 1$) . Such properties include the completeness, orthogonality, Plancherel/Parseval, periodicity, shift, convolution, and unitarity properties above, as well as many FFT algorithms. For this reason, the discrete Fourier transform can be defined by using roots of unity in fields other than the complex numbers, and such generalizations are commonly called number-theoretic transforms (NTTs) in the case of finite fields. For more information, see number-theoretic transform and discrete Fourier transform (general).

Other finite groups

The standard DFT acts on a sequence x0, x1, ..., xN−1 of complex numbers, which can be viewed as a function {0, 1, ..., N − 1} → C. The multidimensional DFT acts on multidimensional sequences, which can be viewed as functions

$$\{0,1,\ldots,N_1-1\}\times\cdots\times\{0,1,\ldots,N_d-1\}\to\mathbb{C}.$$

This suggests the generalization to Fourier transforms on arbitrary finite groups, which act on functions G → C where G is a finite group. In this framework, the standard DFT is seen as the Fourier transform on a cyclic group, while the multidimensional DFT is a Fourier transform on a direct sum of cyclic groups.

Alternatives

For more details on this topic, see Discrete wavelet transform § Comparison with Fourier transform.

There are various alternatives to the DFT for various applications, prominent among which are wavelets. The analog of the DFT is the discrete wavelet transform (DWT). From the point of view of time–frequency analysis, a key limitation of the Fourier transform is that it does not include location information, only frequency information, and thus has difficulty in representing transients. As wavelets have location as well as frequency, they are better able to represent location, at the expense of greater difficulty representing frequency. For details, see comparison of the discrete wavelet transform with the discrete Fourier transform.

DISCRETE-TIME FOURIER TRANSFORM

In mathematics, the discrete-time Fourier transform (DTFT) is a form of Fourier analysis that is applicable to the uniformly-spaced samples of a continuous function. The term discrete-time refers to the fact that the transform operates on discrete data (samples) whose interval often has units of time. From only the samples, it produces a function of frequency that is a periodic summation of the continuous Fourier transform of the original continuous function. Under certain theoretical conditions, described by the sampling theorem, the original continuous function can be recovered perfectly from the DTFT and thus from the original discrete samples. The DTFT itself is a continuous function of frequency, but discrete samples of it can be readily calculated via the discrete Fourier transform (DFT) (see Sampling the DTFT), which is by far the most common method of modern Fourier analysis.

Both transforms are invertible. The inverse DTFT is the original sampled data sequence. The inverse DFT is a periodic summation of the original sequence. The fast Fourier transform (FFT) is an algorithm for computing one cycle of the DFT, and its inverse produces one cycle of the inverse DFT.

The discrete-time Fourier transform of a discrete set of real or complex numbers x[n], for all integers n, is a Fourier series, which produces a periodic function of a frequency variable. When the frequency variable, ω, has normalized units of radians/sample, the periodicity is 2π, and the Fourier series is:

$$X_{2\pi}(\omega) = \sum_{n=-\infty}^{\infty} x[n]\, e^{-i\omega n}.$$

The utility of this frequency domain function is rooted in the Poisson summation formula. Let X(f) be the Fourier transform of any function, x(t), whose samples at some interval T (seconds) are equal (or proportional to) the x[n] sequence, i.e. $T \cdot x(nT) = x[n]$. Then the periodic function represented by the Fourier series is a periodic summation of X(f). In terms of frequency f in hertz (cycles/sec):

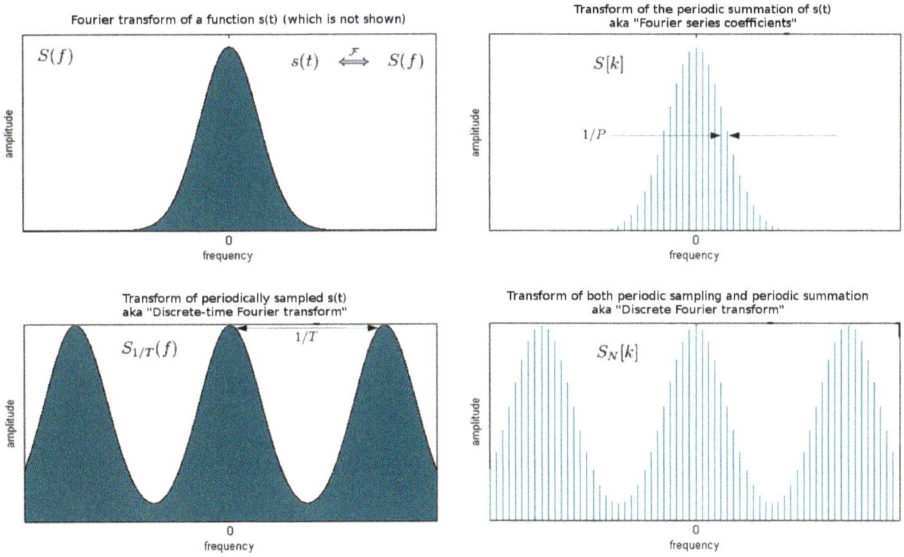

Figure 1. Depiction of a Fourier transform (upper left) and its periodic summation (DTFT) in the lower left corner. The lower right corner depicts samples of the DTFT that are computed by a discrete Fourier transform (DFT).

The integer k has units of cycles/sample, and 1/T is the sample-rate, fs (samples/sec). So X1/T(f) comprises exact copies of X(f) that are shifted by multiples of fs hertz and combined by addition. For sufficiently large fs the k=0 term can be observed in the region [−fs/2, fs/2] with little or no distortion (aliasing) from the other terms. In Fig.1, the extremities of the distribution in the upper left corner are masked by aliasing in the periodic summation (lower left).

We also note that $e^{-i2\pi fTn}$ is the Fourier transform of $\delta(t-nT)$. Therefore, an alternative definition of DTFT is:

$$X_{1/T}(f) = \mathcal{F}\left\{\sum_{n=-\infty}^{\infty} x[n]\cdot\delta(t-nT)\right\}$$

The modulated Dirac comb function is a mathematical abstraction sometimes referred to as impulse sampling.

Inverse Transform

An operation that recovers the discrete data sequence from the DTFT function is called an inverse DTFT. For instance, the inverse continuous Fourier transform of both sides of Eq.3 produces the sequence in the form of a modulated Dirac comb function:

$$\sum_{n=-\infty}^{\infty} x[n]\cdot\delta(t-nT) = \mathcal{F}^{-1}\left\{X_{1/T}(f)\right\} \stackrel{\text{def}}{=} \int_{-\infty}^{\infty} X_{1/T}(f)\cdot e^{i2\pi ft}df.$$

However, noting that X1/T(f) is periodic, all the necessary information is contained within any interval of length 1/T. In both Eq.1 and Eq.2, the summations over n are a

Fourier series, with coefficients x[n]. The standard formulas for the Fourier coefficients are also the inverse transforms:

$$x[n] = T\int_{\frac{1}{T}} X_{1/T}(f)\cdot e^{i2\pi fnT}df \quad \text{(integral over any interval of length } 1/T\text{)}$$
$$= \frac{1}{2\pi}\int_{2\pi} X_{2\pi}(\omega)\cdot e^{i\omega n}d\omega \quad \text{(integral over any interval of length } 2\pi\text{)}$$

Periodic Data

When the input data sequence x[n] is N-periodic, Eq.2 can be computationally reduced to a discrete Fourier transform (DFT), because:

- All the available information is contained within N samples.
- $X_{1/T}(f)$ converges to zero everywhere except integer multiples of $\frac{1}{NT}$ known as harmonic frequencies.
- The DTFT is periodic, so the maximum number of unique harmonic amplitudes is $\frac{1}{T}/\frac{1}{NT}=N$.

The kernel $x[n]\cdot e^{-i2\pi fTn}$ is N-periodic at the harmonic frequencies, $f=\frac{k}{NT}$ Introducing the notation $\sum_N$ to represent a sum over any n-sequence of length N, we can write:

$$X_{1/T}\left(\frac{k}{NT}\right) = \sum_{m=-\infty}^{\infty}\left(\sum_N x[n-mN]\cdot e^{-i2\pi\frac{k}{N}(n-mN)}\right)$$
$$= \sum_{m=-\infty}^{\infty}\left(\sum_N x[n]\cdot e^{-i2\pi\frac{k}{N}n}\right) = T\underbrace{\left(\sum_N x(nT)\cdot e^{-i2\pi\frac{k}{N}n}\right)}_{X[k]\quad\text{(DFT)}}\cdot\left(\sum_{m=-\infty}^{\infty}1\right).$$

Therefore, the DTFT diverges at the harmonic frequencies, but at different frequency-dependent rates. And those rates are given by the DFT of one cycle of the x[n] sequence. In terms of a Dirac comb function, this is represented by:

$$X_{1/T}(f) = \sum_{k=-\infty}^{\infty}(T\cdot X[k])\cdot\frac{1}{NT}\delta\left(f-\frac{k}{NT}\right) = \underbrace{\frac{1}{N}\sum_{k=-\infty}^{\infty}X[k]\cdot\delta\left(f-\frac{k}{NT}\right)}_{\text{DTFT of a periodic sequence}}.$$

Sampling the DTFT

When the DTFT is continuous, a common practice is to compute an arbitrary number of samples (N) of one cycle of the periodic function $X_{1/T}$:

$$\underbrace{X_{1/T}\left(\frac{k}{NT}\right)}_{X_k} = \sum_{n=-\infty}^{\infty}x[n]\cdot e^{-i2\pi\frac{kn}{N}} \qquad k=0,\dots,N-1$$
$$= \underbrace{\sum_N x_N[n]\cdot e^{-i2\pi\frac{kn}{N}}}_{DFT}, \quad \text{(sum over any } n\text{-sequence of length } N\text{)}$$

where xN is a periodic summation:

$$x_N[n] \stackrel{\text{def}}{=} \sum_{m=-\infty}^{\infty} x[n-mN].$$

The xN sequence is the inverse DFT. Thus, our sampling of the DTFT causes the inverse transform to become periodic. The array of $|X_k|^2$ is known as a periodogram, where parameter N is known to Matlab users as NFFT.

In order to evaluate one cycle of xN numerically, we require a finite-length x[n] sequence. For instance, a long sequence might be truncated by a window function of length L resulting in two cases worthy of special mention: L ≤ N and L = I•N, for some integer I (typically 6 or 8). For notational simplicity, consider the x[n] values below to represent the modified values.

When L = I•N a cycle of xN reduces to a summation of I blocks of length N. This goes by various names, such as:

- window-presum FFT
- Weight, OverLap, Add (WOLA)
- polyphase FFT
- multiple block windowing
- time-aliasing.

An interesting way to understand/motivate the technique is to recall that decimation of sampled data in one domain (time or frequency) produces aliasing in the other, and vice versa. The xN summation is mathematically equivalent to aliasing, leading to decimation in frequency, leaving only DTFT samples least affected by spectral leakage. That is usually a priority when implementing an FFT filter-bank (channelizer). With a conventional window function of length L, scalloping loss would be unacceptable. So multi-block windows are created using FIR filter design tools. Their frequency profile is flat at the highest point and falls off quickly at the midpoint between the remaining DTFT samples. The larger the value of parameter I the better the potential performance. We note that the same results can be obtained by computing and decimating an L-length DFT, but that is not computationally efficient.

When L ≤ N the DFT is usually written in this more familiar form:

$$X_k = \sum_{n=0}^{N-1} x[n] \cdot e^{-i2\pi \frac{kn}{N}}.$$

In order to take advantage of a fast Fourier transform algorithm for computing the DFT, the summation is usually performed over all N terms, even though N-L of them are zeros. Therefore, the case L < N is often referred to as "zero-padding".

Spectral leakage, which increases as L decreases, is detrimental to certain important performance metrics, such as resolution of multiple frequency components and the amount of noise measured by each DTFT sample. But those things don't always matter, for instance when the x[n] sequence is a noiseless sinusoid (or a constant), shaped by a window function. Then it is a common practice to use zero-padding to graphically display and compare the detailed leakage patterns of window functions. To illustrate that for a rectangular window, consider the sequence:

$$x[n] = e^{i2\pi \frac{1}{8} n}, \quad \text{and } L = 64.$$

The two figures below are plots of the magnitude of two different sized DFTs, as indicated in their labels. In both cases, the dominant component is at the signal frequency: f = 1/8 = 0.125. Also visible on the right is the spectral leakage pattern of the L = 64 rectangular window. The illusion on the left is a result of sampling the DTFT at all of its zero-crossings. Rather than the DTFT of a finite-length sequence, it gives the impression of an infinitely long sinusoidal sequence. Contributing factors to the illusion are the use of a rectangular window, and the choice of a frequency (1/8 = 8/64) with exactly 8 (an integer) cycles per 64 samples.

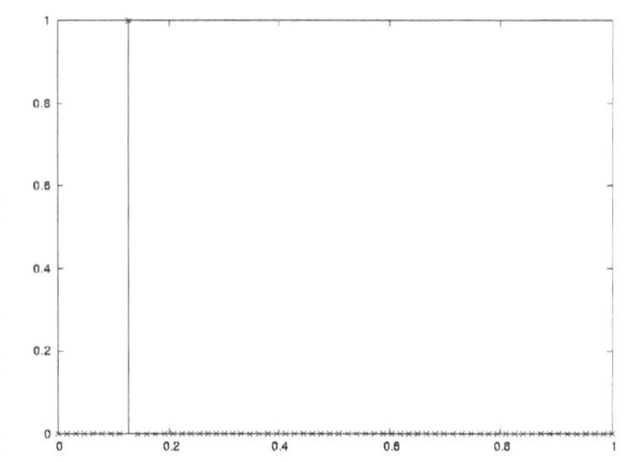

DFT for L = 64 and N = 64

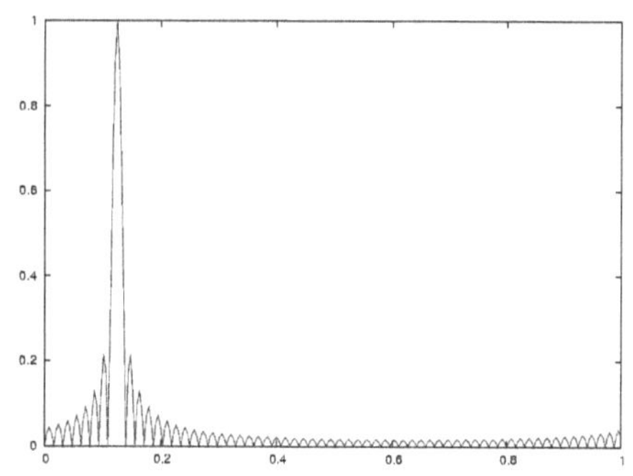

DFT for L = 64 and N = 256

Convolution

The convolution theorem for sequences is:

$$x * y = \mathrm{DTFT}^{-1}\left[\mathrm{DTFT}\{x\} \cdot \mathrm{DTFT}\{y\}\right].$$

An important special case is the circular convolution of sequences x and y defined by xN * y where xN is a periodic summation. The discrete-frequency nature of DTFT{xN} "selects" only discrete values from the continuous function DTFT{y}, which results in considerable simplification of the inverse transform. As shown at Convolution theorem#Functions of discrete variable sequences:

$$x_N * y = \mathrm{DTFT}^{-1}\left[\mathrm{DTFT}\{x_N\} \cdot \mathrm{DTFT}\{y\}\right] = \mathrm{DFT}^{-1}\left[\mathrm{DFT}\{x_N\} \cdot \mathrm{DFT}\{y_N\}\right].$$

For x and y sequences whose non-zero duration is less than or equal to N, a final simplification is:

$$x_N * y = \mathrm{DFT}^{-1}\left[\mathrm{DFT}\{x\} \cdot \mathrm{DFT}\{y\}\right].$$

The significance of this result is expounded at Circular convolution and Fast convolution algorithms.

Relationship to the Z-transform

The bilateral Z-transform is defined by:

$$\underbrace{\mathcal{Z}\{x[n]\}}_{\widehat{X}(z)} = \sum_{n=-\infty}^{\infty} x[n]\, z^{-n},$$

where z is a complex variable.

We denote this function as $\widehat{X}(z)$ to avoid confusion with the Fourier transform. For the special case that z is constrained to values of the form {\displaystyle e^{i\omega }:}

$$\widehat{X}(e^{i\omega}) = \sum_{n=-\infty}^{\infty} x[n]\, e^{-i\omega n} = X_{2\pi}(\omega) = X_{1/T}\left(\tfrac{\omega}{2\pi T}\right) = \sum_{k=-\infty}^{\infty} X\left(\tfrac{\omega}{2\pi T} - k/T\right)$$
$$= \sum_{k=-\infty}^{\infty} X\left(\tfrac{\omega-2\pi k}{2\pi T}\right)$$

Note that when parameter T changes, the terms of {\displaystyle X_{2\pi }(\omega)} remain a constant separation (2π) apart, and their width scales up or down. The terms of {\displaystyle X_{1/T}(f)} remain a constant width and their separation (1/T) scales up or down.

Table of discrete-time Fourier transforms

Some common transform pairs are shown in the table below. The following notation applies:

- ω = 2πfT is a real number representing continuous angular frequency (in radians per sample). (f is in cycles/sec, and T is in sec/sample.) In all cases in the table, the DTFT is 2π-periodic (in ω).
- $X_{2\pi}(\omega)$ designates a function defined on $-\infty < \omega < \infty$.
- $X_o(\omega)$ designates a function defined on $-\pi < \omega \leq \pi$, and zero elsewhere. Then: $e^{i\omega}$

$$\widehat{X}(e^{i\omega}) = \sum_{n=-\infty}^{\infty} x[n]\, e^{-i\omega n} = X_{2\pi}(\omega) = X_{1/T}\left(\tfrac{\omega}{2\pi T}\right) = \sum_{k=-\infty}^{\infty} X\left(\tfrac{\omega}{2\pi T} - k/T\right)$$
$$= \sum_{k=-\infty}^{\infty} X\left(\tfrac{\omega-2\pi k}{2\pi T}\right)$$

- δ(ω) is the Dirac delta function
- sinc(t) is the normalized sinc function
- rect(t) is the rectangle function
- tri(t) is the triangle function
- n is an integer representing the discrete-time domain (in samples)
- u[n] is the discrete-time unit step function
- δ[n] is the Kronecker delta $\delta_{n,0}$

Time domain x[n]	Frequency domain $X_{2\pi}(\omega)$	Remarks
$\delta[n]$	$X_{2\pi}(\omega) = 1$	
$\delta[n-M]$	$X_{2\pi}(\omega) = e^{-i\omega M}$	integer M
$\sum_{m=-\infty}^{\infty} \delta[n-Mm]$	$X_{2\pi}(\omega) = \sum_{m=-\infty}^{\infty} e^{-i\omega Mm} = \frac{2\pi}{M} \sum_{k=-\infty}^{\infty} \delta\left(\omega - \frac{2\pi k}{M}\right)$ $X_o(\omega) = \frac{2\pi}{M} \sum_{k=-(M-1)/2}^{(M-1)/2} \delta\left(\omega - \frac{2\pi k}{M}\right)$ odd M $X_o(\omega) = \frac{2\pi}{M} \sum_{k=-M/2+1}^{M/2} \delta\left(\omega - \frac{2\pi k}{M}\right)$ even M	integer $M > 0$
$u[n]$	$X_{2\pi}(\omega) = \frac{1}{1-e^{-i\omega}} + \pi \sum_{k=-\infty}^{\infty} \delta(\omega - 2\pi k)$ $X_o(\omega) = \frac{1}{1-e^{-i\omega}} + \pi \cdot \delta(\omega)$	The $1/(1-e^{-i\omega})$ term must be interpreted as a distribution in the sense of a Cauchy principal value around its poles at ω = 2πk.
$a^n u[n]$	$X_{2\pi}(\omega) = \frac{1}{1-ae^{-i\omega}}$	$0 < \|a\| < 1$
e^{-ian}	$X_o(\omega) = 2\pi \cdot \delta(\omega + a)$, -p = a < p $X_{2\pi}(\omega) = 2\pi \sum_{k=-\infty}^{\infty} \delta(\omega + a - 2\pi k)$	real number a
$\cos(an)$	$X_o(\omega) = \pi[\delta(\omega + a) + \delta(\omega - a)]$, -p < a < p $X_{2\pi}(\omega) = \pi \sum_{k=-\infty}^{\infty} [\delta(\omega - a - 2\pi k) + \delta(\omega + a + 2\pi k)]$	real number a
$\sin(an)$	$X_{2\pi}(\omega) = \frac{\pi}{i} \sum_{k=-\infty}^{\infty} [\delta(\omega - a - 2\pi k) - \delta(\omega + a + 2\pi k)]$	real number a
$\operatorname{rect}\left[\frac{n - M/2}{M}\right]$	$X_o(\omega) = \frac{\sin[\omega(M+1)/2]}{\sin(\omega/2)} e^{-\frac{i\omega M}{2}}$	integer M
$\operatorname{sinc}(a+n)$	$X_o(\omega) = e^{ia\omega}$	real number a
$W \cdot \operatorname{sinc}^2(Wn)$	$X_o(\omega) = \operatorname{tri}\left(\frac{\omega}{2\pi W}\right)$	real number W 0 < W = 0.5
$W \cdot \operatorname{sinc}(Wn)$	$X_o(\omega) = \operatorname{rect}\left(\frac{\omega}{2\pi W}\right)$	real numbers W 0 < W = 1
$\begin{cases} 0 & n = 0 \\ \frac{(-1)^n}{n} & \text{elsewhere} \end{cases}$	$X_o(\omega) = j\omega$	it works as a differentiator filter
$\frac{W}{(n+a)} \{\cos[\pi W(n+a)] - \operatorname{sinc}[W(n+a)]\}$	$X_o(\omega) = j\omega \cdot \operatorname{rect}\left(\frac{\omega}{\pi W}\right) e^{ja\omega}$	real numbers W, a 0 < W = 1
$\begin{cases} \frac{\pi}{2} & n = 0 \\ \frac{(-1)^n - 1}{\pi n^2} & \text{otherwise} \end{cases}$	$X_o(\omega) = \|\omega\|$	
$\begin{cases} 0; & n \text{ even} \\ \frac{2}{\pi n}; & n \text{ odd} \end{cases}$	$X_o(\omega) = \begin{cases} j & \omega < 0 \\ 0 & \omega = 0 \\ -j & \omega > 0 \end{cases}$	Hilbert transform
$\frac{C(A+B)}{2\pi} \cdot \operatorname{sinc}\left[\frac{A-B}{2\pi} n\right] \cdot \operatorname{sinc}\left[\frac{A+B}{2\pi} n\right]$	$X_o(\omega) =$ 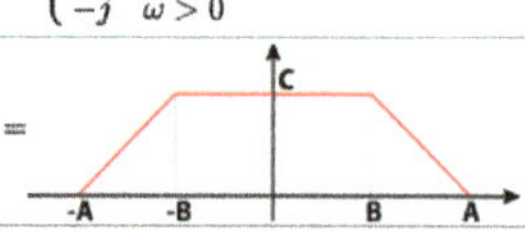	real numbers A, B complex C

Properties

This table shows some mathematical operations in the time domain and the corresponding effects in the frequency domain.

- {\displaystyle *\!} is the discrete convolution of two sequences
- x[n]* is the complex conjugate of x[n]

Property	Time domain $x[n]$	Frequency domain $X_{2\pi}(\omega)$	Remarks
Linearity	$a \cdot x[n] + b \cdot y[n]$	$a \cdot X_{2\pi}(\omega) + b \cdot Y_{2\pi}(\omega)$	
Shift in time	$x[n-k]$	$X_{2\pi}(\omega) \cdot e^{-i\omega k}$	integer *k*
Shift in frequency	$x[n] \cdot e^{ian}$	$X_{2\pi}(\omega - a)$	real number *a*
down-sampling	$x[nM]$	$\frac{1}{M}\sum_{m=0}^{M-1} X_{2\pi}\left(\frac{\omega-2\pi m}{M}\right)$	integer *M*
Expansion	$\begin{cases} x[n/M] & n=\text{multiple of M} \\ 0 & \text{otherwise} \end{cases}$	$\sum_{m=0}^{M-1} X_o\left(M\left(\omega - \frac{2\pi m}{M}\right)\right)$	integer *M*
Time reversal	$x[-n]$	$X_{2\pi}(-\omega)$	
Time conjugation	$x[n]^*$	$X_{2\pi}(-\omega)^*$	
Time reversal & conjugation	$x[-n]^*$	$X_{2\pi}(\omega)^*$	
Derivative in frequency	$\frac{n}{i}x[n]$	$\frac{dX_{2\pi}(\omega)}{d\omega}$	
Integral in frequency	$\frac{i}{n}x[n]$	$\int_{-\pi}^{\omega} X_{2\pi}(\nu)d\nu$	
Convolve in time	$x[n] * y[n]$	$X_{2\pi}(\omega) \cdot Y_{2\pi}(\omega)$	
Multiply in time	$x[n] \cdot y[n]$	$\frac{1}{2\pi}\int_{-\pi}^{\pi} X_{2\pi}(\nu) \cdot Y_{2\pi}(\omega - \nu)d\nu$	Periodic convolution
Cross correlation	$\rho_{xy}[n] = x[-n]^* * y[n]$	$R_{xy}(\omega) = X_{2\pi}(\omega)^* \cdot Y_{2\pi}(\omega)$	
Parseval's theorem	$E = \sum_{n=-\infty}^{\infty} x[n] \cdot y[n]^*$	$E = \frac{1}{2\pi}\int_{-\pi}^{\pi} X_{2\pi}(\omega) \cdot Y_{2\pi}(\omega)^* d\omega$	

IMPULSE INVARIANCE

Impulse invariance is a technique for designing discrete-time infinite-impulse-response (IIR) filters from continuous-time filters in which the impulse response of the continuous-time system is sampled to produce the impulse response of the discrete-time system. The frequency response of the discrete-time system will be a sum of shifted copies of the frequency response of the continuous-time system; if the continuous-time system is approximately band-limited to a frequency less than the Nyquist frequency of the sampling, then the frequency response of the discrete-time system will be approximately equal to it for frequencies below the Nyquist frequency.

Discussion

The continuous-time system's impulse response, $h_c(t)$, is sampled with sampling period T to produce the discrete-time system's impulse response, $h[n]$.

$$h[n] = Th_c(nT)$$

Thus, the frequency responses of the two systems are related by

$$H(e^{j\omega}) = \sum_{k=-\infty}^{\infty} H_c\left(j\frac{\omega}{T} + j\frac{2\pi}{T}k\right)$$

If the continuous time filter is approximately band-limited. (i.e. $H_c(j\Omega) < \delta$ when $|\Omega| \geq \pi/T$), then the frequency response of the discrete-time system will be approximately the continuous-time system's frequency response for frequencies below π radians per sample (below the Nyquist frequency 1/(2T) Hz):

$H(e^{j\omega}) = H_c(j\omega/T)$ for $|\omega| \leq \pi$

Comparison to the Bilinear transform

Note that aliasing will occur, including aliasing below the Nyquist frequency to the extent that the continuous-time filter's response is nonzero above that frequency. The bilinear transform is an alternative to impulse invariance that uses a different mapping that maps the continuous-time system's frequency response, out to infinite frequency, into the range of frequencies up to the Nyquist frequency in the discrete-time case, as opposed to mapping frequencies linearly with circular overlap as impulse invariance does.

Effect on Poles in System Function

If the continuous poles at , the system function can be written in partial fraction expansion as

$$H_c(s) = \sum_{k=1}^{N} \frac{A_k}{s - s_k}$$

Thus, using the inverse Laplace transform, the impulse response is

$$h_c(t) = \begin{cases} \sum_{k=1}^{N} A_k e^{s_k t}, & t \geq 0 \\ 0, & \text{otherwise} \end{cases}$$

The corresponding discrete-time system's impulse response is then defined as the following

$$h[n] = Th_c(nT)$$

$$h[n] = T\sum_{k=1}^{N} A_k e^{s_k n T} u[n]$$

Performing a z-transform on the discrete-time impulse response produces the following discrete-time system function

$$H(z) = T\sum_{k=1}^{N} \frac{A_k}{1 - e^{s_k T} z^{-1}}$$

Thus the poles from the continuous-time system function are translated to poles at $z = e^{s_k T}$. The zeros, if any, are not so simply mapped.

Poles and Zeros

If the system function has zeros as well as poles, they can be mapped the same way, but the result is no longer an impulse invariance result: the discrete-time impulse response is not equal simply to samples of the continuous-time impulse response. This method is known as the matched Z-transform method, or pole–zero mapping. In the case of all-pole filters, the methods are equivalent.

Stability and Causality

Since poles in the continuous-time system at s = sk transform to poles in the discrete-time system at z = exp(skT), poles in the left half of the s-plane map to inside the unit circle in the z-plane; so if the continuous-time filter is causal and stable, then the discrete-time filter will be causal and stable as well.

Corrected Formula

When a causal continuous-time impulse response has a discontinuity at t=0, the expressions above are not consistent. This is because $h_c(0)$ should really only contribute half its value to $h[0]$.

Making this correction gives

$$h[n] = T\left(h_c(nT) - \frac{1}{2}h_c(0)\delta[n]\right)$$

$$h[n] = T\sum_{k=1}^{N} A_k e^{s_k nT}\left(u[n] - \frac{1}{2}\delta[n]\right)$$

Performing a z-transform on the discrete-time impulse response produces the following discrete-time system function

$$H(z) = T\sum_{k=1}^{N} \frac{A_k}{1 - e^{s_k T} z^{-1}} - \frac{T}{2}\sum_{k=1}^{N} A_k.$$

MATCHED Z-TRANSFORM METHOD

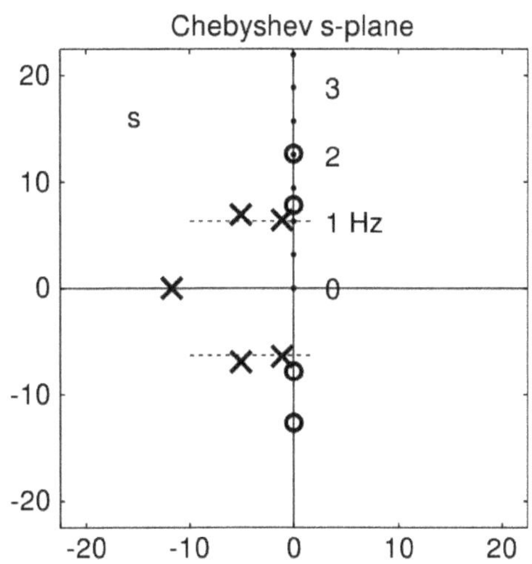

The s-plane poles and zeros of a 5th-order Chebyshev type II lowpass filter to be approximated as a discrete-time filter

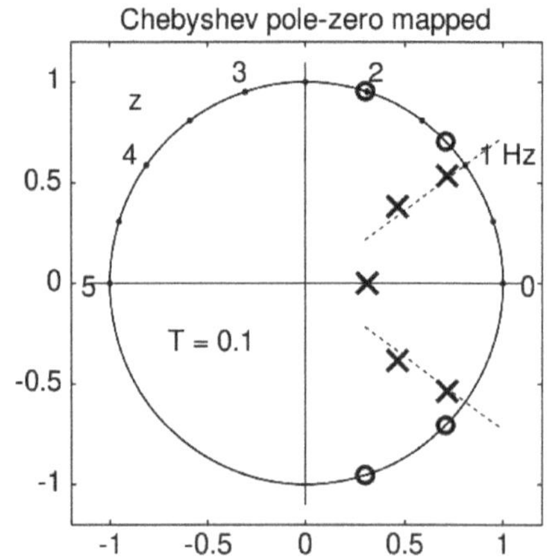

The z-plane poles and zeros of the discrete-time Chebyshev filter, as mapped into the z-plane using the matched Z-transform method with T = 1/10 second. The labeled frequency points and band-edge dotted lines have also been mapped through the function $z=e^{i\omega T}$, to show how frequencies along the $i\omega$ axis in the s-plane map onto the unit circle in the z-plane.

The matched Z-transform method, also called the pole–zero mapping or pole–zero matching method, is a technique for converting a continuous-time filter design to a discrete-time filter (digital filter) design.

The method works by mapping all poles and zeros of the s-plane design to z-plane locations z=esT, for a sample interval T.

Alternative methods include the bilinear transform and impulse invariance methods.

Z-TRANSFORM

For standard z-score in statistics, see Standard score. For Fisher z-transformation in statistics, see Fisher transformation.

In mathematics and signal processing, the Z-transform converts a discrete-time signal, which is a sequence of real or complex numbers, into a complex frequency domain representation.

It can be considered as a discrete-time equivalent of the Laplace transform. This similarity is explored in the theory of time scale calculus.

The basic idea now known as the Z-transform was known to Laplace, and it was re-introduced in 1947 by W. Hurewicz and others as a way to treat sampled-data control systems used with radar. It gives a tractable way to solve linear, constant-coefficient difference equations. It was later dubbed the z-transform by Ragazzini and Zadeh in the sampled-data control group at Columbia University in 1952.

The modified or advanced Z-transform was later developed and popularized by E. I. Jury.

The idea contained within the Z-transform is also known in mathematical literature as the method of generating functions which can be traced back as early as 1730 when it was introduced by de Moivre in conjunction with probability theory. From a mathematical view the Z-transform can also be viewed as a Laurent series where one views the sequence of numbers under consideration as the (Laurent) expansion of an analytic function.

The Z-transform, like many integral transforms, can be defined as either a one-sided or two-sided transform.

Bilateral Z-Transform

The bilateral or two-sided Z-transform of a discrete-time signal x[n] is the formal power series X(z) defined as

$$X(z) = \mathcal{Z}\{x[n]\} = \sum_{n=-\infty}^{\infty} x[n]z^{-n}$$

where n is an integer and z is, in general, a complex number:

$$z = Ae^{j\phi} = A(\cos\phi + j\sin\phi)$$

where A is the magnitude of z, j is the imaginary unit, and ϕ is the complex argument (also referred to as angle or phase) in radians.

Unilateral Z-Transform

Alternatively, in cases where x[n] is defined only for $n \geq 0$, the single-sided or unilateral Z-transform is defined as

$$X(z) = \mathcal{Z}\{x[n]\} = \sum_{n=0}^{\infty} x[n]z^{-n}.$$

In signal processing, this definition can be used to evaluate the Z-transform of the unit impulse response of a discrete-time causal system.

An important example of the unilateral Z-transform is the probability-generating function, where the component x[n] is the probability that a discrete random variable takes the value n, and the function X(z) is usually written as X(s), in terms of $s = z^{-1}$. The properties of Z-transforms (below) have useful interpretations in the context of probability theory.

Geophysical definition

In geophysics, the usual definition for the Z-transform is a power series in z as opposed to z^{-1}. This convention is used, for example, by Robinson and Treitel and by Kanasewich. The geophysical definition is:

$$X(z) = \mathcal{Z}\{x[n]\} = \sum_{n} x[n]z^{n}.$$

The two definitions are equivalent; however, the difference results in a number of changes. For example, the location of zeros and poles move from inside the unit circle using one definition, to outside the unit circle using the other definition. Thus, care is required to note which definition is being used by a particular author.

Inverse Z-transform

The inverse Z-transform is

$$x[n] = \mathcal{Z}^{-1}\{X(z)\} = \frac{1}{2\pi j}\oint_C X(z)z^{n-1}dz$$

where C is a counterclockwise closed path encircling the origin and entirely in the region of convergence (ROC). In the case where the ROC is causal (see Example 2), this means the path C must encircle all of the poles of X(z).

A special case of this contour integral occurs when C is the unit circle (and can be used when the ROC includes the unit circle which is always guaranteed when X(z) is stable, i.e. all the poles are within the unit circle). The inverse Z-transform simplifies to the inverse discrete-time Fourier transform:

$$x[n] = \frac{1}{2\pi}\int_{-\pi}^{+\pi} X(e^{j\omega})e^{j\omega n}\,d\omega.$$

The Z-transform with a finite range of n and a finite number of uniformly spaced z values can be computed efficiently via Bluestein's FFT algorithm. The discrete-time Fourier transform (DTFT)—not to be confused with the discrete Fourier transform (DFT)—is a special case of such a Z-transform obtained by restricting z to lie on the unit circle.

Region of Convergence

The region of convergence (ROC) is the set of points in the complex plane for which the Z-transform summation converges.

$$ROC = \left\{ z : \left| \sum_{n=-\infty}^{\infty} x[n]z^{-n} \right| < \infty \right\}$$

Example 1 (no ROC)

Let $x[n] = (0.5)^n$. Expanding $x[n]$ on the interval $(-\infty, \infty)$ it becomes

$$x[n] = \{\cdots, 0.5^{-3}, 0.5^{-2}, 0.5^{-1}, 1, 0.5, 0.5^2, 0.5^3, \cdots\} = \{\cdots, 2^3, 2^2, 2, 1, 0.5, 0.5^2, 0.5^3, \cdots\}.$$

Looking at the sum

$$\sum_{n=-\infty}^{\infty} x[n]z^{-n} \to \infty.$$

Therefore, there are no values of z that satisfy this condition.

Example 2 (causal ROC)

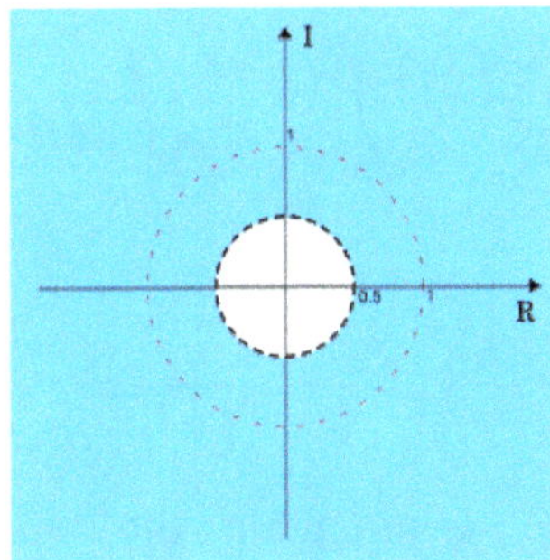

ROC shown in blue, the unit circle as a dotted grey circle (appears reddish to the eye) and the circle |z| = 0.5 is shown as a dashed black circle

Let $x[n] = 0.5^n u[n]$ (where u is the Heaviside step function). Expanding x[n] on the interval $(-\infty, \infty)$ it becomes

$x[n] = \{\cdots, 0, 0, 0, 1, 0.5, 0.5^2, 0.5^3, \cdots\}$.

Looking at the sum

$$\sum_{n=-\infty}^{\infty} x[n]z^{-n} = \sum_{n=0}^{\infty} 0.5^n z^{-n} = \sum_{n=0}^{\infty} \left(\frac{0.5}{z}\right)^n = \frac{1}{1-0.5z^{-1}}$$

The last equality arises from the infinite geometric series and the equality only holds if $|0.5z^{-1}| < 1$ which can be rewritten in terms of z as $|z| > 0.5$. Thus, the ROC is $|z| > 0.5$. In this case the ROC is the complex plane with a disc of radius 0.5 at the origin "punched out".

Example 3 (anticausal ROC)

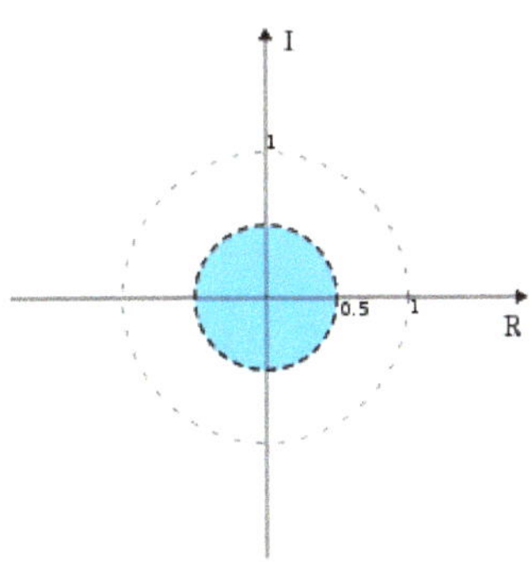

ROC shown in blue, the unit circle as a dotted grey circle and the circle |z| = 0.5 is shown as a dashed black circle

Let $x[n] = 0.5^n u[n]$ (where u is the Heaviside step function). Expanding x[n] on the interval (−∞, ∞) it becomes

$$\sum_{n=-\infty}^{\infty} x[n]z^{-n} = \sum_{n=0}^{\infty} 0.5^n z^{-n} = \sum_{n=0}^{\infty}\left(\frac{0.5}{z}\right)^n = \frac{1}{1-0.5z^{-1}}.$$

Looking at the sum

$$\sum_{n=-\infty}^{\infty} x[n]z^{-n} = \sum_{n=0}^{\infty} 0.5^n z^{-n} = \sum_{n=0}^{\infty}\left(\frac{0.5}{z}\right)^n = \frac{1}{1-0.5z^{-1}}.$$

Using the infinite geometric series, again, the equality only holds if $|0.5-1z| < 1$ which can be rewritten in terms of z as $|z| < 0.5$. Thus, the ROC is $|z| < 0.5$. In this case the ROC is a disc centered at the origin and of radius 0.5.

What differentiates this example from the previous example is only the ROC. This is intentional to demonstrate that the transform result alone is insufficient.

Examples conclusion

Examples 2 & 3 clearly show that the Z-transform X(z) of x[n] is unique when and only when specifying the ROC. Creating the pole–zero plot for the causal and anticausal case show that the ROC for either case does not include the pole that is at 0.5. This extends to cases with multiple poles: the ROC will never contain poles.

In example 2, the causal system yields an ROC that includes $|z| = \infty$ while the anticausal system in example 3 yields an ROC that includes $|z| = 0$.

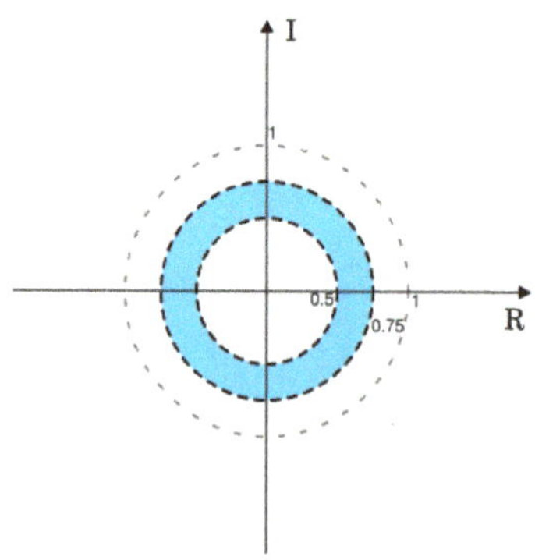

ROC shown as a blue ring $0.5 < |z| < 0.75$

In systems with multiple poles it is possible to have an ROC that includes neither $|z| = \infty$ nor $|z| = 0$. The ROC creates a circular band. For example,

$$x[n] = 0.5^n u[n] - 0.75^n u[-n-1]$$

has poles at 0.5 and 0.75. The ROC will be $0.5 < |z| < 0.75$, which includes neither the origin nor infinity. Such a system is called a mixed-causality system as it contains a causal term $(0.5)^n u[n]$ and an anticausal term $-(0.75)^n u[-n-1]$.

The stability of a system can also be determined by knowing the ROC alone. If the ROC contains the unit circle (i.e., $|z| = 1$) then the system is stable. In the above systems the causal system (Example 2) is stable because $|z| > 0.5$ contains the unit circle.

If you are provided a Z-transform of a system without an ROC (i.e., an ambiguous x[n]) you can determine a unique x[n] provided you desire the following:

- Stability
- Causality

If you need stability then the ROC must contain the unit circle. If you need a causal system then the ROC must contain infinity and the system function will be a right-sided sequence. If you need an anticausal system then the ROC must contain the origin and the system function will be a left-sided sequence. If you need both, stability and causality, all the poles of the system function must be inside the unit circle.

The unique x[n] can then be found.

Properties

	Time domain	Z-domain	Proof	ROC
Notation	$x[n] = \mathcal{Z}^{-1}\{X(z)\}$	$X(z) = \mathcal{Z}\{x[n]\}$		$r_2 < \|z\| < r_1$
Linearity	$a_1 x_1[n] + a_2 x_2[n]$	$a_1 X_1(z) + a_2 X_2(z)$	$X(z) = \sum_{n=-\infty}^{\infty} (a_1 x_1(n) + a_2 x_2(n)) z^{-n}$ $= a_1 \sum_{n=-\infty}^{\infty} x_1(n) z^{-n} + a_2 \sum_{n=-\infty}^{\infty} x_2(n) z^{-n}$ $= a_1 X_1(z) + a_2 X_2(z)$	Contains $ROC_1 \cap ROC_2$
Time expansion	$x_K[n] = \begin{cases} x[r], & n = Kr \\ 0, & n \notin K\mathbb{Z} \end{cases}$ with $K\mathbb{Z} := \{Kr : r \in \mathbb{Z}\}$	$X(z^K)$	$X_K(z) = \sum_{n=-\infty}^{\infty} x_K(n) z^{-n}$ $= \sum_{r=-\infty}^{\infty} x(r) z^{-rK}$ $= \sum_{r=-\infty}^{\infty} x(r)(z^K)^{-r}$ $= X(z^K)$	$R^{\frac{1}{K}}$
Decimation	$x[Kn]$	$\frac{1}{K} \sum_{p=0}^{K-1} X\left(z^{\frac{1}{K}} \cdot e^{-i\frac{2\pi}{K}p}\right)$	ohio-state.edu or ee.ic.ac.uk	
Time shifting	$x[n-k]$	$z^{-k} X(z)$	$Z\{x[n-k]\} = \sum_{n=0}^{\infty} x[n-k] z^{-n}$ $= \sum_{j=-k}^{\infty} x[j] z^{-(j+k)} \quad j = n-k$ $= \sum_{j=-k}^{\infty} x[j] z^{-j} z^{-k}$ $= z^{-k} \sum_{j=-k}^{\infty} x[j] z^{-j}$ $= z^{-k} \sum_{j=0}^{\infty} x[j] z^{-j} \quad x[\beta] = 0, \beta < 0$ $= z^{-k} X(z)$	ROC, except $z = 0$ if $k > 0$ and $z = \infty$ if $k < 0$
Scaling in the z-domain	$a^n x[n]$	$X(a^{-1} z)$	$\mathcal{Z}\{a^n x[n]\} = \sum_{n=-\infty}^{\infty} a^n x(n) z^{-n}$ $= \sum_{n=-\infty}^{\infty} x(n)(a^{-1} z)^{-n}$ $= X(a^{-1} z)$	$\|a\| r_2 < \|z\| < \|a\| r_1$
Time reversal	$x[-n]$	$X(z^{-1})$	$\mathcal{Z}\{x(-n)\} = \sum_{n=-\infty}^{\infty} x(-n) z^{-n}$ $= \sum_{m=-\infty}^{\infty} x(m) z^{m}$ $= \sum_{m=-\infty}^{\infty} x(m)(z^{-1})^{-m}$ $= X(z^{-1})$	$\frac{1}{r_1} < \|z\| < \frac{1}{r_2}$

Complex conjugation	$x^*[n]$	$X^*(z^*)$	$\begin{aligned} \mathcal{Z}\{x^*(n)\} &= \sum_{n=-\infty}^{\infty} x^*(n)z^{-n} \\ &= \sum_{n=-\infty}^{\infty} [x(n)(z^*)^{-n}]^* \\ &= \left[\sum_{n=-\infty}^{\infty} x(n)(z^*)^{-n}\right]^* \\ &= X^*(z^*) \end{aligned}$	
Real part	$\operatorname{Re}\{x[n]\}$	$\frac{1}{2}[X(z)+X^*(z^*)]$		
Imaginary part	$\operatorname{Im}\{x[n]\}$	$\frac{1}{2j}[X(z)-X^*(z^*)]$		
Differentiation	$nx[n]$	$-z\frac{dX(z)}{dz}$	$\begin{aligned} \mathcal{Z}\{nx(n)\} &= \sum_{n=-\infty}^{\infty} nx(n)z^{-n} \\ &= z\sum_{n=-\infty}^{\infty} nx(n)z^{-n-1} \\ &= -z\sum_{n=-\infty}^{\infty} x(n)(-nz^{-n-1}) \\ &= -z\sum_{n=-\infty}^{\infty} x(n)\frac{d}{dz}(z^{-n}) \\ &= -z\frac{dX(z)}{dz} \end{aligned}$	ROC, if $X(z)$ is rational; ROC possibly excluding the boundary, if $X(z)$ is irrational[9]
Convolution	$x_1[n] * x_2[n]$	$X_1(z)X_2(z)$	$\begin{aligned} \mathcal{Z}\{x_1(n) * x_2(n)\} &= \mathcal{Z}\left\{\sum_{l=-\infty}^{\infty} x_1(l)x_2(n-l)\right\} \\ &= \sum_{n=-\infty}^{\infty}\left[\sum_{l=-\infty}^{\infty} x_1(l)x_2(n-l)\right]z^{-n} \\ &= \sum_{l=-\infty}^{\infty} x_1(l)\left[\sum_{n=-\infty}^{\infty} x_2(n-l)z^{-n}\right] \\ &= \left[\sum_{l=-\infty}^{\infty} x_1(l)z^{-l}\right]\left[\sum_{n=-\infty}^{\infty} x_2(n)z^{-n}\right] \\ &= X_1(z)X_2(z) \end{aligned}$	Contains $ROC_1 \cap ROC_2$
Cross-correlation	$r_{x_1,x_2} = x_1^*[-n] * x_2[n]$	$R_{x_1,x_2}(z) = X_1^*(\frac{1}{z^*})X_2(z)$		Contains the intersection of ROC of $X_1(\frac{1}{z^*})$ and $X_2(z)$
First difference	$x[n] - x[n-1]$	$(1-z^{-1})X(z)$		Contains the intersection of ROC of $X_1(z)$ and $z \neq 0$
Accumulation	$\sum_{k=-\infty}^{n} x[k]$	$\frac{1}{1-z^{-1}}X(z)$	$\begin{aligned} \sum_{n=-\infty}^{\infty}\sum_{k=-\infty}^{n} x[k]z^{-n} &= \sum_{n=-\infty}^{\infty}(x[n]+\cdots+x[-\infty])z^{-n} \\ &= X[z]\left(1+z^{-1}+z^{-2}+\cdots\right) \\ &= X[z]\sum_{j=0}^{\infty} z^{-j} \\ &= X[z]\frac{1}{1-z^{-1}} \end{aligned}$	
Multiplication	$x_1[n]x_2[n]$	$\frac{1}{j2\pi}\oint_C X_1(v)X_2(\frac{z}{v})v^{-1}\mathrm{d}v$		-

Parseval's theorem

$$\sum_{n=-\infty}^{\infty} x_1[n]x_2^*[n] \quad = \quad \frac{1}{j2\pi}\oint_C X_1(v)X_2^*(\tfrac{1}{v^*})v^{-1}\mathrm{d}v$$

Initial value theorem: If x[n] is causal, then

$$x[0] = \lim_{z\to\infty} X(z).$$

Final value theorem: If the poles of (z−1)X(z) are inside the unit circle, then

$$x[\infty] = \lim_{z\to 1}(z-1)X(z).$$

Table of common Z-Transform Pairs

Here:

$$u : n \mapsto u[n] = \begin{cases} 1, & n \geq 0 \\ 0, & n < 0 \end{cases}$$

is the unit (or Heaviside) step function and

$$\delta : n \mapsto \delta[n] = \begin{cases} 1, & n = 0 \\ 0, & n \neq 0 \end{cases}$$

is the discrete-time unit impulse function (cf Dirac delta function which is a continuous-time version). The two functions are chosen together so that the unit step function is the accumulation (running total) of the unit impulse function.

	Signal, $x[n]$	Z-transform, $X(z)$	ROC
1	$\delta[n]$	1	all z
2	$\delta[n-n_0]$	z^{-n_0}	$z \neq 0$
3	$u[n]$	$\frac{1}{1-z^{-1}}$	$\lvert z\rvert > 1$
4	$e^{-\alpha n}u[n]$	$\frac{1}{1-e^{-\alpha}z^{-1}}$	$\lvert z\rvert > e^{-\alpha}$
5	$-u[n-1]$	$\frac{1}{1-z}$	$\lvert z\rvert < 1$
6	$nu[n]$	$\frac{z^{-1}}{(1-z^{-1})^2}$	$\lvert z\rvert > 1$
7	$-nu[-n-1]$	$\frac{z^{-1}}{(1-z^{-1})^2}$	$\lvert z\rvert < 1$
8	$n^2u[n]$	$\frac{z^{-1}(1+z^{-1})}{(1-z^{-1})^3}$	$\lvert z\rvert > 1$
9	$-n^2u[-n-1]$	$\frac{z^{-1}(1+z^{-1})}{(1-z^{-1})^3}$	$\lvert z\rvert < 1$
10	$n^3u[n]$	$\frac{z^{-1}(1+4z^{-1}+z^{-2})}{(1-z^{-1})^4}$	$\lvert z\rvert > 1$
11	$-n^3u[-n-1]$	$\frac{z^{-1}(1+4z^{-1}+z^{-2})}{(1-z^{-1})^4}$	$\lvert z\rvert < 1$
12	$a^nu[n]$	$\frac{1}{1-az^{-1}}$	$\lvert z\rvert > \lvert a\rvert$
13	$-a^nu[-n-1]$	$\frac{1}{1-az^{-1}}$	$\lvert z\rvert < \lvert a\rvert$
14	$na^nu[n]$	$\frac{az^{-1}}{(1-az^{-1})^2}$	$\lvert z\rvert > \lvert a\rvert$
15	$-na^nu[-n-1]$	$\frac{az^{-1}}{(1-az^{-1})^2}$	$\lvert z\rvert < \lvert a\rvert$

16	$n^2 a^n u[n]$	$\frac{az^{-1}(1+az^{-1})}{(1-az^{-1})^3}$	$\|z\| > \|a\|$
17	$-n^2 a^n u[-n-1]$	$\frac{az^{-1}(1+az^{-1})}{(1-az^{-1})^3}$	$\|z\| < \|a\|$
18	$\binom{n+m-1}{m-1} a^n u[n]$	$\frac{1}{(1-az^{-1})^m}$, for positive integer m	$\|z\| > \|a\|$
19	$(-1)^m \binom{-n-1}{m-1} a^n u[-n-m]$	$\frac{1}{(1-az^{-1})^m}$, for positive integer m	$\|z\| < \|a\|$
20	$\cos(\omega_0 n)u[n]$	$\frac{1-z^{-1}\cos(\omega_0)}{1-2z^{-1}\cos(\omega_0)+z^{-2}}$	$\|z\| > 1$
21	$\sin(\omega_0 n)u[n]$	$\frac{z^{-1}\sin(\omega_0)}{1-2z^{-1}\cos(\omega_0)+z^{-2}}$	$\|z\| > 1$
22	$a^n \cos(\omega_0 n)u[n]$	$\frac{1-az^{-1}\cos(\omega_0)}{1-2az^{-1}\cos(\omega_0)+a^2z^{-2}}$	$\|z\| > \|a\|$
23	$a^n \sin(\omega_0 n)u[n]$	$\frac{az^{-1}\sin(\omega_0)}{1-2az^{-1}\cos(\omega_0)+a^2z^{-2}}$	$\|z\| > \|a\|$

Relationship to Fourier Series and Fourier transform

For values of z in the region |z|=1, known as the unit circle, we can express the transform as a function of a single, real variable, ω, by defining $z=e^{j\omega}$. And the bi-lateral transform reduces to a Fourier series:

$$\sum_{n=-\infty}^{\infty} x[n]\, z^{-n} = \sum_{n=-\infty}^{\infty} x[n]\, e^{-j\omega n},$$

which is also known as the discrete-time Fourier transform (DTFT) of the x[n] sequence. This 2π-periodic function is the periodic summation of a Fourier transform, which makes it a widely used analysis tool. To understand this, let X(f) be the Fourier transform of any function, x(t), whose samples at some interval, T, equal the x[n] sequence. Then the DTFT of the x[n] sequence can be written as:

$$\underbrace{\sum_{n=-\infty}^{\infty} \overbrace{x(nT)}^{x[n]}\, e^{-j2\pi f nT}}_{\text{DTFT}} = \frac{1}{T}\sum_{k=-\infty}^{\infty} X(f-k/T).$$

When T has units of seconds, f has units of hertz. Comparison of the two series reveals that $\omega=2\pi fT$ is a normalized frequency with units of radians per sample. The value ω=2π corresponds to ω=2π corresponds to $f=\frac{1}{T}$ Hz. And now, with the substitution

$f = \frac{\omega}{2\pi T}$, Eq.1 can be expressed in terms of the Fourier transform, X(•):

$$\sum_{n=-\infty}^{\infty} x[n]\, e^{-j\omega n} = \frac{1}{T} \sum_{k=-\infty}^{\infty} \underbrace{X\left(\frac{\omega}{2\pi T} - \frac{k}{T}\right)}_{X\left(\frac{\omega - 2\pi k}{2\pi T}\right)}.$$

When sequence x(nT) represents the impulse response of an LTI system, these functions are also known as its frequency response. When the x(nT) sequence is periodic, its DTFT is divergent at one or more harmonic frequencies, and zero at all other frequencies. This is often represented by the use of amplitude-variant Dirac delta functions at the harmonic frequencies. Due to periodicity, there are only a finite number of unique amplitudes, which are readily computed by the much simpler discrete Fourier transform (DFT). (See DTFT; periodic data.)

Relationship to Laplace transform

Bilinear Transform

The bilinear transform can be used to convert continuous-time filters (represented in the Laplace domain) into discrete-time filters (represented in the Z-domain), and vice versa. The following substitution is used:

$$s = \frac{2}{T}\frac{(z-1)}{(z+1)}$$

to convert some function $H(s)$ in the Laplace domain to a function $H(z)$ in the Z-domain (Tustin transformation), or

$$z = \frac{2 + sT}{2 - sT}$$

from the Z-domain to the Laplace domain. Through the bilinear transformation, the complex s-plane (of the Laplace transform) is mapped to the complex z-plane (of the z-transform). While this mapping is (necessarily) nonlinear, it is useful in that it maps the entire jΩ axis of the s-plane onto the unit circle in the z-plane. As such, the Fourier transform (which is the Laplace transform evaluated on the jΩ axis) becomes the discrete-time Fourier transform. This assumes that the Fourier transform exists; i.e., that the jΩ axis is in the region of convergence of the Laplace transform.

Starred Transform

Given a one-sided Z-transform, X(z), of a time-sampled function, the corresponding starred transform produces a Laplace transform and restores the dependence on sampling parameter, T:

$$X^*(s) = X(z)\Big|_{z = e^{sT}}$$

The inverse Laplace transform is a mathematical abstraction known as an impulse-sampled function.

Linear constant-coefficient difference equation

The linear constant-coefficient difference (LCCD) equation is a representation for a linear system based on the autoregressive moving-average equation.

$$\sum_{p=0}^{N} y[n-p]\alpha_p = \sum_{q=0}^{M} x[n-q]\beta_q$$

Both sides of the above equation can be divided by α0, if it is not zero, normalizing α0 = 1 and the LCCD equation can be written

$$Y(z)\sum_{p=0}^{N} z^{-p}\alpha_p = X(z)\sum_{q=0}^{M} z^{-q}\beta_q$$

This form of the LCCD equation is favorable to make it more explicit that the "current" output y[n] is a function of past outputs y[n−p], current input x[n], and previous inputs x[n−q].

Transfer Function

Taking the Z-transform of the above equation (using linearity and time-shifting laws) yields

$$Y(z)\sum_{p=0}^{N} z^{-p}\alpha_p = X(z)\sum_{q=0}^{M} z^{-q}\beta_q$$

$$H(z) = \frac{Y(z)}{X(z)} = \frac{\sum_{q=0}^{M} z^{-q}\beta_q}{\sum_{p=0}^{N} z^{-p}\alpha_p} = \frac{\beta_0 + z^{-1}\beta_1 + z^{-2}\beta_2 + \cdots + z^{-M}\beta_M}{\alpha_0 + z^{-1}\alpha_1 + z^{-2}\alpha_2 + \cdots + z^{-N}\alpha_N}.$$

Zeros and poles

From the fundamental theorem of algebra the numerator has M roots (corresponding to zeros of H) and the denominator has N roots (corresponding to poles). Rewriting the transfer function in terms of poles and zeros

$$H(z) = \frac{(1-q_1 z^{-1})(1-q_2 z^{-1})\cdots(1-q_M z^{-1})}{(1-p_1 z^{-1})(1-p_2 z^{-1})\cdots(1-p_N z^{-1})}$$

where qk is the k-th zero and pk is the k-th pole. The zeros and poles are commonly complex and when plotted on the complex plane (z-plane) it is called the pole–zero plot.

In addition, there may also exist zeros and poles at z = 0 and z = ∞. If we take these poles and zeros as well as multiple-order zeros and poles into consideration, the number of zeros and poles are always equal.

By factoring the denominator, partial fraction decomposition can be used, which can then be transformed back to the time domain. Doing so would result in the impulse response and the linear constant coefficient difference equation of the system.

Output response

If such a system H(z) is driven by a signal X(z) then the output is Y(z) = H(z)X(z). By performing partial fraction decomposition on Y(z) and then taking the inverse Z-transform the output y[n] can be found. In practice, it is often useful to fractionally decompose $\frac{Y(z)}{z}$ before multiplying that quantity by z to generate a form of Y(z) which has terms with easily computable inverse Z-transforms.

ZAK TRANSFORM

In mathematics, the Zak transform is a certain operation which takes as input a function of one variable and produces as output a function of two variables. The output function is called the Zak transform of the input function. The transform is defined as an infinite series in which each term is a product of a dilation of a translation by an integer of the function and an exponential function. In applications of Zak transform to signal processing the input function represents a signal and the transform will be a mixed time–frequency representation of the signal. The signal may be real valued or complex-valued, defined on a continuous set (for example, the real numbers) or a discrete set (for example, the integers or a finite subset of integers). The Zak transform is a generalization of the discrete Fourier transform.

The Zak transform had been discovered by several people in different fields and was called by different names. It was called the Gel'fand mapping because I.M. Gel'fand introduced it in his work on eigenfunction expansions. The transform was rediscovered independently by J. Zak in 1967 who called it the "k-q representation". There seems to be a general consent among experts in the field to call it the Zak transform, since Zak was indeed the first to systematically study that transform in a more general setting and recognize its usefulness.

Continuous-time Zak transform: Definition

In defining the continuous-time Zak transform, the input function is a function of a real variable. So, let f(t) be a function of a real variable t. The continuous-time Zak transform of f(t) is a function of two real variables one of which is t. The other variable may be denoted by w. The continuous-time Zak transform has been defined variously.

Definition 1

Let a be a positive constant. The Zak transform of f(t), denoted by Za[f], is a function of t and w defined by

$$Z_a[f](t,w) = \sqrt{a} \sum_{k=-\infty}^{\infty} f(at+ak)e^{-2\pi kwi}.$$

Definition 2

The special case of Definition 1 obtained by taking a = 1 is sometimes taken as the definition of the Zak transform. In this special case, the Zak transform of f(t) is denoted by Z[f].

$$Z[f](t,w) = \sum_{k=-\infty}^{\infty} f(t+k)e^{-2\pi kwi}.$$

Definition 3

The notation Z[f] is used to denote another form of the Zak transform. In this form, the Zak transform of f(t) is defined as follows:

$$Z[f](t,\nu) = \sum_{k=-\infty}^{\infty} f(t+k)e^{-k\nu i}.$$

Definition 4

Let T be a positive constant. The Zak transform of f(t), denoted by ZT[f], is a function of t and w defined by

$$Z_T[f](t,w) = \sqrt{T} \sum_{k=-\infty}^{\infty} f(t+kT)e^{-2\pi kwTi}.$$

Here t and w are assumed to satisfy the conditions $0 \le t \le T$ and $0 \le w \le 1/T$.

Example

The Zak transform of the function

$$\phi(t) = \begin{cases} 1, & 0 \le t < 1 \\ 0, & \text{otherwise} \end{cases}$$

is given by

$$Z[\phi](t,w) = e^{-2\pi\lceil -t\rceil wi}$$

where $\lceil -t\rceil$ denotes the smallest integer not less than -t (the ceil function).

Properties of the Zak transform

In the following it will be assumed that the Zak transform is as given in Definition 2.

1. Linearity

Let a and b be any real or complex numbers. Then

$$Z[af+bg](t,w)=aZ[f](t,w)+bZ[g](t,w)$$

2. Periodicity

$$Z[f](t,w+1)=Z[f](t,w)$$

3. Quasi-periodicity

$$Z[f](t+1,w)=e^{2\pi wi}Z[f](t,w)$$

4. Conjugation

$$Z[\bar{f}](t,w)=\overline{Z[f]}(t,-w)$$

5. Symmetry

If $f(t)$ is even then $Z[f](t,w)=Z[f](-t,-w)$

If $f(t)$ is odd then $Z[f](t,w)=-Z[f](-t,-w)$

6. Convolution

Let * denote convolution with respect to the variable t.

$$Z[f\star g](t,w)=Z[f](t,w)\star Z[g](t,w)$$

Inversion Formula

Given the Zak transform of a function, the function can be reconstructed using the following formula:

$$f(t)=\int_0^1 Z[f](t,w)\,dw.$$

Discrete Zak transform: Definition

In defining the discrete Zak transform, the input function is a function of an integer variable. So, let f(n) be a function of an integer variable n (n taking all positive, zero and negative integers as values). The discrete Zak transform of f(n) is a function of two real variables one of which is the integer variable n. The other variable is a real variable which may be denoted by w. The discrete Zak transform has also been defined variously. However, only one of the definitions is given below.

Definition

The discrete Zak transform of the function f(n) where n is an integer variable, denoted by Z[f], is defined by

$$Z[f](n,w) = \sum_{k=-\infty}^{\infty} f(n+k)e^{-2\pi kwi}.$$

Inversion formula

Given the discrete transform of a function f(n), the function can be reconstructed using the following formula:

$$f(n) = \int_0^1 Z[f](n,w)\,dw.$$

Applications

The Zak transform has been used successfully used in physics in quantum field theory, in electrical engineering in time-frequency representation of signals, and in digital data transmission. The Zak transform has also applications in mathematics. For example, it has been used in the Gabor representation problem.

REFERENCES

1. Brigham, E. Oran (1988). The fast Fourier transform and its applications. Englewood Cliffs, N.J.: Prentice Hall. ISBN 0-13-307505-2.
2. Oppenheim, Alan V.; Schafer, R. W.; and Buck, J. R. (1999). Discrete-time signal processing. Upper Saddle River, N.J.: Prentice Hall. ISBN 0-13-754920-2.
3. Smith, Steven W. (1999). "Chapter 8: The Discrete Fourier Transform". The Scientist and Engineer's Guide to Digital Signal Processing (Second ed.). San Diego, Calif.: California Technical Publishing. ISBN 0-9660176-3-3.
4. Cormen, Thomas H.; Charles E. Leiserson; Ronald L. Rivest; Clifford Stein (2001). "Chapter 30: Polynomials and the FFT". Introduction to Algorithms (Second ed.). MIT Press and McGraw-Hill. pp. 822–848. ISBN 0-262-03293-7. esp. section 30.2: The DFT and FFT, pp. 830–838.
5. P. Duhamel; B. Piron; J. M. Etcheto (1988). On computing the inverse DFT. IEEE Trans. Acoust., Speech and Sig. Processing. **36** (2): 285–286. doi:10.1109/29.1519.
6. J. H. McClellan; T. W. Parks (1972). Eigenvalues and eigenvectors of the discrete Fourier transformation. IEEE Trans. Audio Electroacoust. **20** (1): 66–74. doi:10.1109/TAU.1972.1162342.
7. Bradley W. Dickinson; Kenneth Steiglitz (1982). Eigenvectors and functions of the discrete Fourier transform. IEEE Trans. Acoust., Speech and Sig. Processing. **30** (1): 25–31. doi:10.1109/TASSP.1982.1163843. (Note that this paper has an apparent typo in its table of the eigenvalue multiplicities: the +i/−i columns are interchanged. The correct table can be found in McClellan and Parks, 1972, and is easily confirmed numerically.)

8. F. A. Grünbaum (1982). The eigenvectors of the discrete Fourier transform. J. Math. Anal. Appl. **88** (2): 355–363. doi:10.1016/0022-247X(82)90199-8.

9. Natig M. Atakishiyev; Kurt Bernardo Wolf (1997). Fractional Fourier-Kravchuk transform. J. Opt. Soc. Am. A. **14** (7): 1467–1477. Bibcode:1997JOSAA..14.1467A. doi:10.1364/JOSAA.14.001467.

10. C. Candan; M. A. Kutay; H. M.Ozaktas (2000). The discrete fractional Fourier transform. IEEE Trans. on Signal Processing. **48** (5): 1329–1337. Bibcode:2000ITSP...48.1329C. doi:10.1109/78.839980.

11. Magdy Tawfik Hanna, Nabila Philip Attalla Seif, and Waleed Abd El Maguid Ahmed (2004). Hermite-Gaussian-like eigenvectors of the discrete Fourier transform matrix based on the singular-value decomposition of its orthogonal projection matrices. IEEE Trans. Circ. Syst. I. **51** (11): 2245–2254. doi:10.1109/TCSI.2004.836850.

12. Shamgar Gurevich; Ronny Hadani (2009). On the diagonalization of the discrete Fourier transform. Applied and Computational Harmonic Analysis. **27** (1): 87–99. arXiv:0808.3281. doi:10.1016/j.acha.2008.11.003. preprint at.

13. Shamgar Gurevich; Ronny Hadani; Nir Sochen (2008). The finite harmonic oscillator and its applications to sequences, communication and radar. IEEE Transactions on Information Theory. **54** (9): 4239–4253. arXiv:0808.1495. doi:10.1109/TIT.2008.926440. preprint at.

14. Juan G. Vargas-Rubio; Balu Santhanam (2005). On the multiangle centered discrete fractional Fourier transform. IEEE Sig. Proc. Lett. **12** (4): 273–276. Bibcode:2005ISPL...12..273V. doi:10.1109/LSP.2005.843762.

15. J. Cooley, P. Lewis, and P. Welch (1969). The finite Fourier transform. IEEE Trans. Audio Electroacoustics. **17** (2): 77–85. doi:10.1109/TAU.1969.1162036.

16. F.N. Kong (2008). Analytic Expressions of Two Discrete Hermite-Gaussian Signals. IEEE Trans. Circuits and Systems –II: Express Briefs. **55** (1): 56–60. doi:10.1109/TCSII.2007.909865.

17. E. R. Kanasewich (1981). Time sequence analysis in geophysics (3rd ed.). University of Alberta. pp. 185–186. ISBN 978-0-88864-074-1.

18. J. R. Ragazzini and L. A. Zadeh (1952). The analysis of sampled-data systems. Trans. Am. Inst. Elec. Eng. **71** (II): 225–234.

19. Cornelius T. Leondes (1996). Digital control systems implementation and computational techniques. Academic Press. p. 123. ISBN 978-0-12-012779-5.

20. Eliahu Ibrahim Jury (1958). Sampled-Data Control Systems. John Wiley & Sons.

21. Eliahu Ibrahim Jury (1973). Theory and Application of the Z-Transform Method. Krieger Pub Co. ISBN 0-88275-122-0.

22. Eliahu Ibrahim Jury (1964). Theory and Application of the Z-Transform Method. John Wiley & Sons. p. 1.

23. Enders A. Robinson, Sven Treitel (2008). Digital Imaging and Deconvolution: The ABCs of Seismic Exploration and Processing Digital Imaging and Deconvolution: The ABCs of Seismic Exploration and Processing. SEG Books. pp. 163, 375–376. ISBN 9781560801481.

24. E. R. Kanasewich (1981). Time Sequence Analysis in Geophysics. University of Alberta. pp. 186, 249. ISBN 9780888640741.

25. R. Forouzan (2016). Region of convergence of derivative of Z transform. Elec. Lett. IET. **52** (8): 617–619. doi:10.1049/el.2016.0189.

CHAPTER 5

Theory of Sampling

Chapter Contents

- **Introduction**
- **Aliasing**
- **Description**
- **Bandpass Signals**
- **Sampling Sinusoidal Functions**
- **Anti-Aliasing Filter**
- **Decimation (Signal Processing)**

The signals we use in the real world, such as our voices, are called "analog" signals. To process these signals in computers, we need to convert the signals to "digital" form. While an analog signal is continuous in both time and amplitude, a digital signal is discrete in both time and amplitude. To convert a signal from continuous time to discrete time, a process called sampling is used. The value of the signal is measured at certain intervals in time. Each measurement is referred to as a sample. (The analog signal is also quantized in amplitude, but that process is ignored in this demonstration.

When the continuous analog signal is sampled at a frequency F, the resulting discrete signal has more frequency components than did the analog signal. To be precise, the frequency components of the analog signal are repeated at the sample rate. That is, in the discrete frequency response they are seen at their original position, and are also seen centered around +/- F, and around +/- 2F, etc.

How many samples are necessary to ensure we are preserving the information contained in the signal? If the signal contains high frequency components, we will need to sample at a higher rate to avoid losing information that is in the signal. In general, to preserve the full information in the signal, it is necessary to sample at twice the maximum frequency of the signal. This is known as the Nyquist rate. The Sampling Theorem states that a signal can be exactly reproduced if it is sampled at a frequency F, where F is greater than twice the maximum frequency in the signal.

INTRODUCTION

In signal processing, sampling is the reduction of a continuous-time signal to a discrete-time signal. A common example is the conversion of a sound wave (a continuous signal) to a sequence of samples (a discrete-time signal).

A sample is a value or set of values at a point in time and/or space.

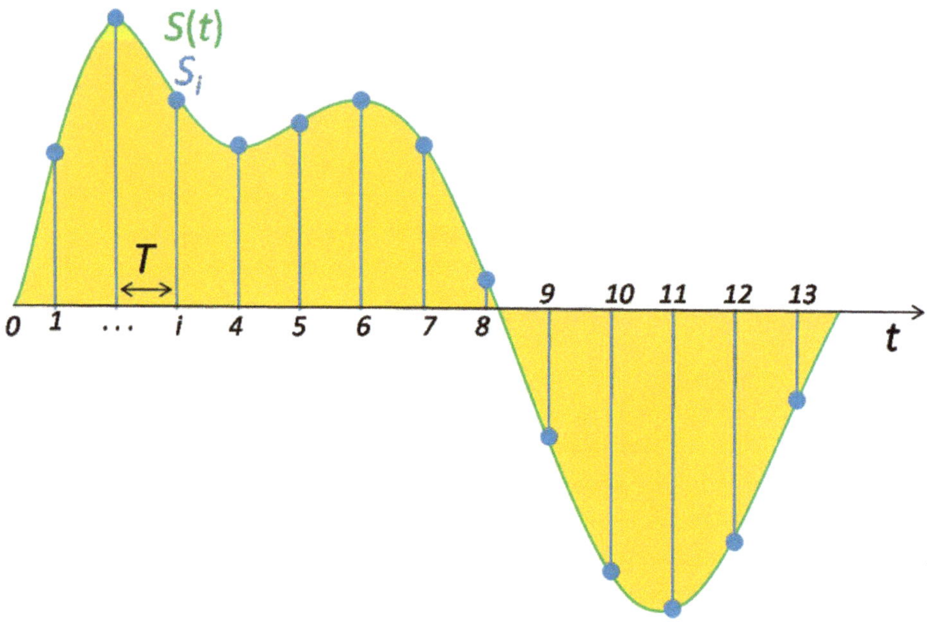

Signal sampling representation. The continuous signal is represented with a green colored line while the discrete samples are indicated by the blue vertical lines.

A sampler is a subsystem or operation that extracts samples from a continuous signal.

A theoretical ideal sampler produces samples equivalent to the instantaneous value of the continuous signal at the desired points.

ALIASING

In signal processing and related disciplines, aliasing is an effect that causes different signals to become indistinguishable (or aliases of one another) when sampled. It also refers to the distortion or artifact that results when the signal reconstructed from samples is different from the original continuous signal.

Aliasing can occur in signals sampled in time, for instance digital audio, and is referred to as temporal aliasing. Aliasing can also occur in spatially sampled signals, for instance moiré patterns in digital images. Aliasing in spatially sampled signals is called spatial aliasing.

Aliasing is generally avoided by applying low pass filters- anti-aliasing filters to the analog signal before sampling.

DESCRIPTION

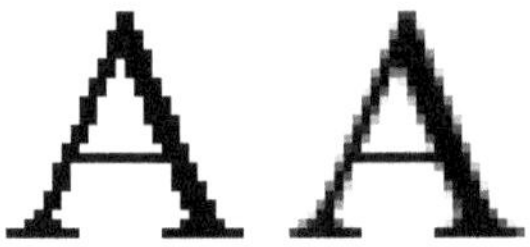

Left: An aliased image of the letter A in Times New Roman. Right: An anti-aliased image. (See also: Font rasterization)

When a digital image is viewed, a reconstruction is performed by a display or printer device, and by the eyes and the brain. If the image data is processed in some ways during sampling or reconstruction, the reconstructed image will differ from the original image, and an alias is seen.

An example of spatial aliasing is the moiré pattern observed in a poorly pixelized image of a brick wall. Spatial anti-aliasing techniques avoid such poor pixelizations. Aliasing can be caused either by the sampling stage or the reconstruction stage; these may be distinguished by calling sampling aliasing prealiasing and reconstruction aliasing postaliasing.

Temporal aliasing is a major concern in the sampling of video and audio signals. Music, for instance, may contain high-frequency components that are inaudible to humans. If a piece of music is sampled at 32000 samples per second (Hz), any frequency components above 16000 Hz (the Nyquist frequency for this sampling rate) will cause aliasing when the music is reproduced by a digital to analog converter (DAC). To prevent this an anti-aliasing filter is used to remove components above the Nyquist frequency prior to sampling.

In video or cinematography, temporal aliasing results from the limited frame rate, and causes the wagon-wheel effect, whereby a spoked wheel appears to rotate too slowly or even backwards. Aliasing has changed its apparent frequency of rotation. A reversal of direction can be described as a negative frequency. Temporal aliasing frequencies in video and cinematography are determined by the frame rate of the camera, but the relative intensity of the aliased frequencies is determined by the shutter timing (exposure time) or the use of a temporal aliasing reduction filter during filming.

Like the video camera, most sampling schemes are periodic; that is, they have a characteristic sampling frequency in time or in space. Digital cameras provide a certain number of samples (pixels) per degree or per radian, or samples per mm in the focal plane of the camera. Audio signals are sampled (digitized) with an analog-to-digital converter, which produces a constant number of samples per second. Some of the most dramatic and subtle examples of aliasing occur when the signal being sampled also has periodic content.

Bandlimited Functions

Main article: Nyquist–Shannon sampling theorem

Actual signals have finite duration and their frequency content, as defined by the Fourier transform, has no upper bound. Some amount of aliasing always occurs when such functions are sampled. Functions whose frequency content is bounded (*bandlimited*) have infinite duration. If sampled at a high enough rate, determined by the *bandwidth*, the original function can in theory be perfectly reconstructed from the infinite set of samples.

BANDPASS SIGNALS

Main article: Undersampling

Sometimes aliasing is used intentionally on signals with no low-frequency content, called bandpass signals. Undersampling, which creates low-frequency aliases, can produce the same result, with less effort, as frequency-shifting the signal to lower frequencies before sampling at the lower rate. Some digital channelizers exploit aliasing in this way for computational efficiency. See Sampling (signal processing), Nyquist rate (relative to sampling), and Filter bank.

SAMPLING SINUSOIDAL FUNCTIONS

Sinusoids are an important type of periodic function, because realistic signals are often modeled as the summation of many sinusoids of different frequencies and different amplitudes (for example, with a Fourier series or transform). Understanding what aliasing does to the individual sinusoids is useful in understanding what happens to their sum.

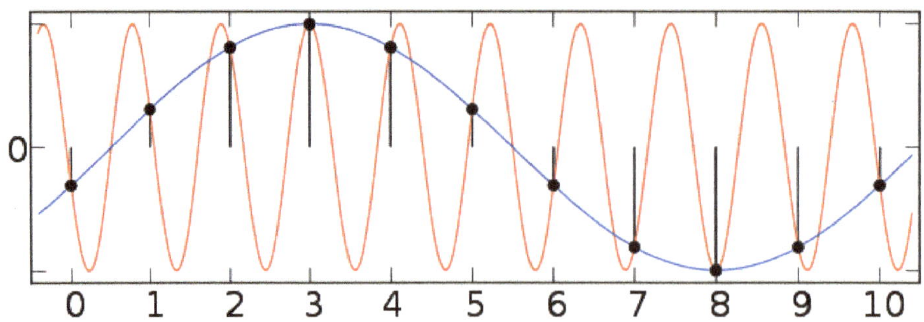

Two different sinusoids that fit the same set of samples.

Here, a plot depicts a set of samples whose sample-interval is 1, and two (of many) different sinusoids that could have produced the samples. The sample-rate in this case is $f_s = 1$. For instance, if the interval is 1 second, the rate is 1 sample per second. Nine cycles of the red sinusoid and 1 cycle of the blue sinusoid span an interval of 10 samples. The corresponding number of cycles per sample are $f_{red} = 0.9$ and $f_{blue} = 0.1$. If these samples were produced by sampling functions $\cos(2\pi(0.9)x-\theta)$ and $\cos(2\pi(0.1)x-\phi)$, they could also have been produced by the trigonometrically identical functions $\cos(2\pi(-0.9)x+\theta)$ and $\cos(2\pi(-0.1)x+\phi)$ which introduces the useful concept of negative frequency.

In general, when a sinusoid of frequency Failed to parse (MathML with SVG or PNG fallback (recommended for modern browsers and accessibility tools): Invalid response ("Math extension cannot connect to Restbase.") from server "/mathoid/local/v1/":): f\, is sampled with frequency f_s the resulting number of cycles per sample is f/f_s (known as normalized frequency), and the samples are indistinguishable from those of another sinusoid (called an alias) whose normalized frequency differs from f/f_s by any integer (positive or negative). Replacing negative frequency sinusoids by their equivalent positive frequency representations, we can express all the aliases of frequency f as f as $f_{alias}(N) \stackrel{\text{def}}{=} |f - Nf_s|$, for any integer N with $f_{alias}(0)=f$ being the true value, and N has units of cycles per sample. Then the $N = 1$ alias of f_{red} is f_{blue}, (and vice versa).

Aliasing matters when one attempts to reconstruct the original waveform from its samples. The most common reconstruction technique produces the smallest of the $f_{alias}(N)$ frequencies. So it is usually important that $f_{alias}(0)$ be the unique minimum. A necessary and sufficient condition for that is $f_s/2 > |f|$, where $f_s/2$ where is commonly called the Nyquist frequency of a system that samples at rate f_s. In our example, the Nyquist condition is satisfied if the original signal is the blue sinusoid ($f = f_{blue}$). But if $f=f_{red}=0.9$, the usual reconstruction method will produce the blue sinusoid instead of the red one.

Folding

In the example above, f_{red} and f_{blue} are symmetrical around the frequency $f_s/2$ And in general, as f increases from 0 to $f_s/2$, $f_{alias}(1)$ decreases from f_s to $f_s/2$. Similarly, as f increases from $f_s/2$ to f_s, falias continues decreasing from N= 1 alias of f_{red} is f_{blue}.

A graph of amplitude vs frequency for a single sinusoid at frequency $0.6f_s$ and some of its aliases at $0.4f_s$, and 1.6fs would look like the 4 black dots in the first figure below. The red lines depict the paths (loci) of the 4 dots if we were to adjust the frequency and amplitude of the sinusoid along the solid red segment (between $f_s/2$ and f_s). No matter what function we choose to change the amplitude vs frequency, the graph will exhibit symmetry between 0 and f_s. This symmetry is commonly referred to as folding, and another name for $f_s/2$ and f_s (the Nyquist frequency) is folding frequency. Folding is often observed in practice when viewing the frequency spectrum of real-valued samples, such as the second figure below.

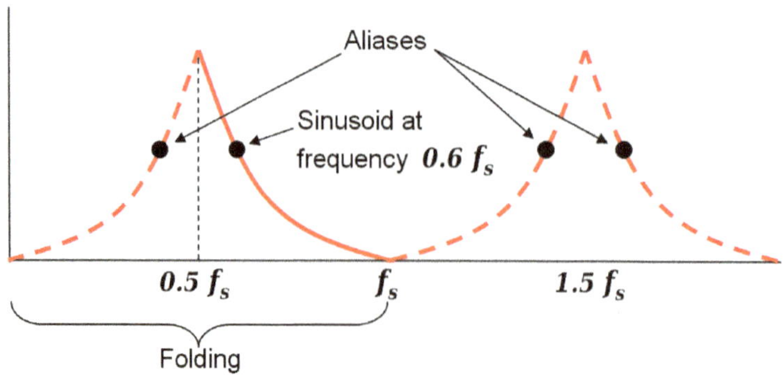

The black dots are aliases of each other. The solid red line is an example of amplitude varying with frequency. The dashed red lines are the corresponding paths of the aliases.

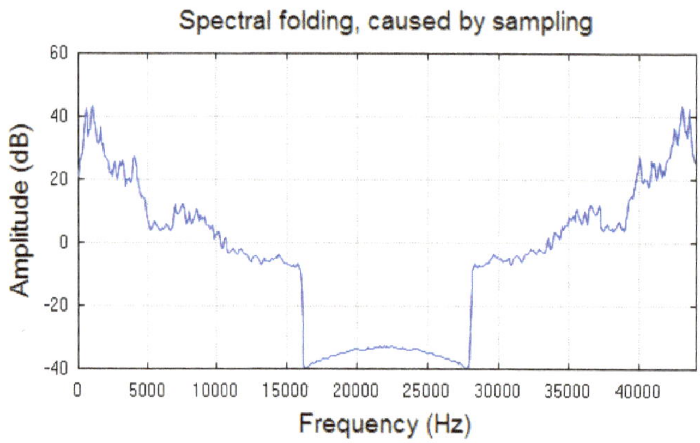

The Fourier transform of music sampled at 44100 samples/sec exhibits symmetry (called "folding") around the Nyquist frequency (22050 Hz).

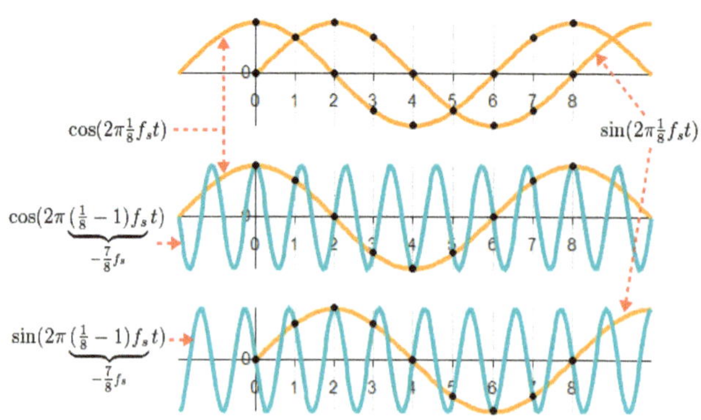

Two complex sinusoids, colored gold and cyan, that fit the same sets of real and imaginary sample points when sampled at the rate (f_s) indicated by the grid lines. The case shown here is: $f_{cyan} = f_{alias}(1) = f_{gold} - 1 \cdot f_s$.

Complex Sinusoids

Complex sinusoids are waveforms whose samples are complex numbers, and the concept of negative frequency is necessary to distinguish them. In that case, the frequencies of the aliases are given by just: $f_{\text{alias}}(N) = f - Nf_s$. Therefore, as $f_s/2$ to f_s, $f_{\text{alias}}(1)$ goes from $-f_s/2$ up to 0. Consequently, complex sinusoids do not exhibit folding. Complex samples of real-valued sinusoids have zero-valued imaginary parts and do exhibit folding.

Sample Frequency

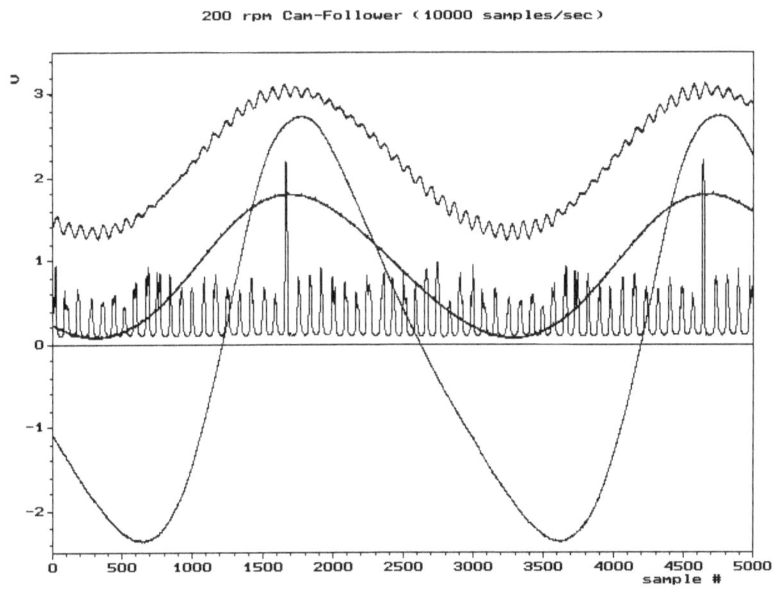

Illustration of 4 waveforms reconstructed from samples taken at six different rates. Two of the waveforms are sufficiently sampled to avoid aliasing at all six rates. The other two illustrate increasing distortion (aliasing) at the lower rates.

When the condition $f_s/2 > f$ is met for the highest frequency component of the original signal, then it is met for all the frequency components, a condition called the Nyquist criterion. That is typically approximated by filtering the original signal to attenuate high frequency components before it is sampled. These attenuated high frequency components still generate low-frequency aliases, but typically at low enough amplitudes that they do not cause problems. A filter chosen in anticipation of a certain sample frequency is called an anti-aliasing filter.

The filtered signal can subsequently be reconstructed, by interpolation algorithms, without significant additional distortion. Most sampled signals are not simply stored and reconstructed. But the fidelity of a theoretical reconstruction (via the Whittaker–Shannon interpolation formula) is a customary measure of the effectiveness of sampling.

Historical Usage

Historically the term aliasing evolved from radio engineering because of the action of superheterodyne receivers. When the receiver shifts multiple signals down to lower frequencies, from RF to IF by heterodyning, an unwanted signal, from an RF frequency equally far from the local oscillator (LO) frequency as the desired signal, but on the wrong side of the LO, can end up at the same IF frequency as the wanted one. If it is strong enough it can interfere with reception of the desired signal. This unwanted signal is known as an image or alias of the desired signal.

Angular Aliasing

Aliasing occurs whenever the use of discrete elements to capture or produce a continuous signal causes frequency ambiguity.

Spatial aliasing, particular of angular frequency, can occur when reproducing a light field or sound field with discrete elements, as in 3D displays or wave field synthesis of sound.

This aliasing is visible in images such as posters with lenticular printing: if they have low angular resolution, then as one moves past them, say from left-to-right, the 2D image does not initially change (so it appears to move left), then as one moves to the next angular image, the image suddenly changes (so it jumps right) – and the frequency and amplitude of this side-to-side movement corresponds to the angular resolution of the image (and, for frequency, the speed of the viewer's lateral movement), which is the angular aliasing of the 4D light field.

The lack of parallax on viewer movement in 2D images and in 3-D film produced by stereoscopic glasses (in 3D films the effect is called "yawing, as the image appears to rotate on its axis) can similarly be seen as loss of angular resolution, all angular frequencies being aliased to 0 (constant).

More Examples

Online Audio Example

The qualitative effects of aliasing can be heard in the following audio demonstration. Six sawtooth waves are played in succession, with the first two sawtooths having a fundamental frequency of 440 Hz (A4), the second two having fundamental frequency of 880 Hz (A5), and the final two at 1760 Hz (A6). The sawtooths alternate between bandlimited (non-aliased) sawtooths and aliased sawtooths and the sampling rate is 22.05 kHz. The bandlimited sawtooths are synthesized from the sawtooth waveform's Fourier series such that no harmonics above the Nyquist frequency are present.

The aliasing distortion in the lower frequencies is increasingly obvious with higher fundamental frequencies, and while the bandlimited sawtooth is still clear at 1760 Hz, the aliased sawtooth is degraded and harsh with a buzzing audible at frequencies lower than the fundamental.

Direction Finding

A form of spatial aliasing can also occur in antenna arrays or microphone arrays used to estimate the direction of arrival of a wave signal, as in geophysical exploration by seismic waves. Waves must be sampled at more than two points per wavelength, or the wave arrival direction becomes ambiguous.

ANTI-ALIASING FILTER

An anti-aliasing filter (AAF) is a filter used before a signal sampler to restrict the bandwidth of a signal to approximately or completely satisfy the sampling theorem over the band of interest. Since the theorem states that unambiguous reconstruction of the signal from its samples is possible when the power of frequencies above the Nyquist frequency is zero, a real anti-aliasing filter trades off between bandwidth and aliasing. A realizable anti-aliasing filter will typically either permit some aliasing to occur or else attenuate some in-band frequencies close to the Nyquist limit. For this reason, many practical systems sample higher than required to ensure that all frequencies of interest can be reconstructed, a practice called oversampling.

Optical Applications

In the case of optical image sampling, as by image sensors in digital cameras, the anti-aliasing filter is also known as an optical low-pass filter, blur filter, or AA filter. The mathematics of sampling in two spatial dimensions is similar to the mathematics of time-domain sampling, but the filter implementation technologies are different. The typical implementation in digital cameras is two layers of birefringent material such as lithium niobate, which spreads each optical point into a cluster of four points.

The choice of spot separation for such a filter involves a tradeoff among sharpness, aliasing, and fill factor (the ratio of the active refracting area of a microlens array to the total contiguous area occupied by the array). In a monochrome or three-CCD or Foveon X3 camera, the microlens array alone, if near 100% effective, can provide a significant anti-aliasing effect, while in color filter array (CFA, e.g. Bayer filter) cameras, an additional filter is generally needed to reduce aliasing to an acceptable level.

The Pentax K-3 from Ricoh introduced a unique sensor-based anti-aliasing filter. The filter works by micro vibrating the sensor element. The user can turn the vibration on or off, selecting anti-aliasing or no anti-aliasing.

Audio Applications

Anti-aliasing filters are commonly used at the input of digital signal processing system's analog to digital converter; similar filters are used as reconstruction filters at the output of such systems, for example in music players. In the latter case, the filter prevents imaging, the reverse process of aliasing where in-band frequencies are mirrored out of band.

Oversampling

A technique known as oversampling is commonly used in audio ADCs. The idea is to use a higher intermediate digital sample rate, so that a nearly-ideal digital filter can sharply cut off aliasing near the original low Nyquist frequency and give better phase response, while a much simpler analog filter can stop frequencies above the new higher Nyquist frequency. Because analog filters have relatively high cost and limited performance, relaxing the demands on the analog filter can greatly reduce both aliasing and cost. Furthermore, because some noise is averaged out, the higher sampling rate can moderately improve SNR.

Alternatively, a signal may be intentionally oversampled without an intermediate frequency to reduce the requirements on the anti-alias filter. For example, CD audio typically extends up to 20 kHz, but is sampled with a 22.05 kHz Nyquist rate. By oversampling by 2.05 kHz, both aliasing and attenuation of higher audio frequencies can be prevented even with less than ideal filters.

In signal processing, oversampling is the process of sampling a signal with a sampling frequency significantly higher than the Nyquist rate. Theoretically a bandwidth-limited signal can be perfectly reconstructed if sampled above the Nyquist rate, which is twice the highest frequency in the signal. Oversampling improves resolution, reduces noise and helps avoid aliasing and phase distortion by relaxing anti-aliasing filter performance requirements.

A signal is said to be oversampled by a factor of N if it is sampled at N times the Nyquist rate.

Motivation

There are three main reasons for performing oversampling:

Anti-aliasing

Oversampling can make it easier to realize analog anti-aliasing filters. Without oversampling, it is very difficult to implement filters with the sharp cutoff necessary to maximize use of the available bandwidth without exceeding the Nyquist limit. By increasing the bandwidth of the sampled signal, design constraints for the anti-aliasing filter may be relaxed. Once sampled, the signal can be digitally filtered and downsampled to the desired sampling frequency. In modern integrated circuit technology, digital filters are easier to implement than comparable analog filters.

Resolution

In practice, oversampling is implemented in order to achieve cheaper higher-resolution A/D and D/A conversion. For instance, to implement a 24-bit converter, it is sufficient to use a 20-bit converter that can run at 256 times the target sampling rate. Combining 256 consecutive 20-bit samples can increase the signal-to-noise ratio at the voltage level by a factor of 16 (the square root of the number of samples averaged), effectively adding 4 bits to the resolution and producing a single sample with 24-bit resolution.

When oversampling by a factor of N, the dynamic range increases by log2(N) bits, because there are N times as many possible values for the sum.

However, the SNR increases by sqrt(N), (not by N as in the article). Summing up uncorrelated noise increases its amplitude by sqrt(N), while summing up a coherent signal increases its average by N. As a result, the SNR (or signal/noise) increases by sqrt(N). In the example, that means while with N=256 there is an increase in Dynamic range by 8 bits, and the content of coherent signal increases by N, but the noise changes by a factor of sqrt(N)=sqrt(256)=16 in the example (not to be confused with an increase of 16 bits), so the SNR changes by a factor of 16.

The number of samples required to get n bits of additional data precision is

$$\text{number of samples} = (2^n)^2 = 2^{2n}.$$

To get the mean sample scaled up to an integer with n additional bits, the sum of 2^{2n} samples is divided by 2^n:

$$\text{scaled mean} = \frac{\sum_{i=0}^{2^{2n}-1} 2^n data_i}{2^{2n}} = \frac{\sum_{i=0}^{2^{2n}-1} data_i}{2^n}.$$

This averaging is only possible if the signal contains equally distributed noise which is enough to be observed by the A/D converter. If not, in the case of a stationary input signal, all 2^n samples would have the same value and the resulting average would be identical to this value; so in this case, oversampling would have made no improvement. (In similar cases where the A/D converter sees no noise and the input signal is changing over time, oversampling still improves the result, but to an inconsistent/unpredictable extent.) This is an interesting counter-intuitive example where adding some dithering noise to the input signal can improve (rather than degrade) the final result because the dither noise allows oversampling to work to improve resolution (or dynamic range). In many practical applications, a small increase in noise is well worth a substantial increase in measurement resolution. In practice, the dithering noise can often be placed outside the frequency range of interest to the measurement, so that this noise can be subsequently filtered out in the digital domain—resulting in a final measurement (in the frequency range of interest) with both higher resolution and lower noise.

Noise

If multiple samples are taken of the same quantity with uncorrelated noise added to each sample, then averaging N samples reduces the noise power by a factor of 1/N. If, for example, we oversample by a factor of 4, the signal-to-noise ratio in terms of power improves by factor of 4 which corresponds to a factor of 2 improvement in terms of voltage.

Certain kinds of A/D converters known as delta-sigma converters produce disproportionately more quantization noise in the upper portion of their output spectrum. By running these converters at some multiple of the target sampling rate, and low-pass filtering the oversampled signal down to half the target sampling rate, a final result with less noise (over the entire band of the converter) can be obtained. Delta-sigma converters use a technique called noise shaping to move the quantization noise to the higher frequencies.

Example

Consider a signal with a bandwidth or highest frequency of B = 100 Hz. The sampling theorem states that sampling frequency would have to be greater than 200 Hz. Sampling at four times that rate requires a sampling frequency of 800 Hz. This gives the anti-aliasing filter a transition band of 300 Hz ((fs/2) − B = (800 Hz/2) − 100 Hz = 300 Hz) instead of 0 Hz if the sampling frequency was 200 Hz.

Achieving an anti-aliasing filter with 0 Hz transition band is unrealistic whereas an anti-aliasing filter with a transition band of 300 Hz is not difficult to create.

Oversampling in Reconstruction

The term oversampling is also used to denote a process used in the reconstruction phase of digital-to-analog conversion, in which an intermediate high sampling rate is used between the digital input and the analogue output. Here, samples are interpolated in the digital domain to add additional samples in between, thereby converting the data to a higher sample rate, which is a form of upsampling. When the resulting higher-rate samples are converted to analog, a less complex/expensive analog low pass filter is required to remove the high-frequency content, which will consist of reflected images of the real signal created by the zero-order hold of the digital-to-analog converter. Essentially, this is a way to shift some of the complexity of the filtering into the digital domain and achieves the same benefit as oversampling in analog-to-digital conversion

Bandpass Signals

Often, an anti-aliasing filter is a low-pass filter; however, this is not a requirement. Generalizations of the Nyquist–Shannon sampling theorem allow sampling of other band-limited passband signals instead of baseband signals.

For signals that are bandwidth limited, but not centered at zero, a band-pass filter can be used as an anti-aliasing filter. For example, this could be done with a single-sideband modulated or frequency modulated signal. If one desired to sample an FM radio broadcast centered at 87.9 MHz and bandlimited to a 200 kHz band, then an appropriate anti-alias filter would be centered on 87.9 MHz with 200 kHz bandwidth (or pass-band of 87.8 MHz to 88.0 MHz), and the sampling rate would be no less than 176.2 MHz, but should also satisfy other constraints to prevent aliasing.

Signal Overload

It is very important to avoid input signal overload when using an anti-aliasing filter. If the signal is strong enough, it can cause clipping at the analog-to-digital converter, even after filtering. When distortion due to clipping occurs after the anti-aliasing filter, it can create components outside the passband of the anti-aliasing filter; these components can then alias, causing the reproduction of other non-harmonically-related frequencies.

DECIMATION (SIGNAL PROCESSING)

In digital signal processing, decimation is the process of reducing the sampling rate of a signal. Complementary to interpolation, which increases sampling rate, it is a specific case of sample rate conversion in a multi-rate digital signal processing system. Decimation utilises filtering to mitigate aliasing distortion, which can occur when simply downsampling a signal. A system component that performs decimation is called a decimator.

In General

Decimation reduces the data rate or the size of the data. The decimation factor is usually an integer or a rational fraction greater than one. This factor multiplies the sampling time or, equivalently, divides the sampling rate. For example, if 16-bit compact disc audio (sampled at 44,100 Hz) is decimated to 22,050 Hz, the audio is said to be decimated by a factor of 2. The bit rate is also reduced in half, from 1,411,200 bit/s to 705,600 bit/s, assuming that each sample retains its bit depth of 16 bits.

By an integer factor

Decimation by an integer factor, M, can be explained as a 2-step process, with an equivalent implementation that is more efficient:

- Reduce high-frequency signal components with a digital lowpass filter.
- Downsample the filtered signal by M; that is, keep only every M^{th} sample.

Downsampling alone causes high-frequency signal components to be misinterpreted by subsequent users of the data, which is a form of distortion called aliasing. The first step, if necessary, is to suppress aliasing to an acceptable level. In this application, the filter is called an anti-aliasing filter, and its design is discussed below. Also see undersampling for information about downsampling bandpass functions and signals.

When the anti-aliasing filter is an IIR design, it relies on feedback from output to input, prior to the downsampling step. With FIR filtering, it is an easy matter to compute only every Mth output. The calculation performed by a decimating FIR filter for the nth output sample is a dot product:

$$y[n] = \sum_{k=0}^{K-1} x[nM - k] \cdot h[k],$$

where the h[•] sequence is the impulse response, and K is its length. x[•] represents the input sequence being downsampled. In a general purpose processor, after computing y[n], the easiest way to compute y[n+1] is to advance the starting index in the x[•] array by M, and recompute the dot product. In the case M=2, h[•] can be designed as a half-band filter, where almost half of the coefficients are zero and need not be included in the dot products.

Impulse response coefficients taken at intervals of M form a subsequence, and there are M such subsequences (phases) multiplexed together. The dot product is the sum of the dot products of each subsequence with the corresponding samples of the x[•] sequence. Furthermore, because of downsampling by M, the stream of x[•] samples involved in any one of the M dot products is never involved in the other dot products. Thus M low-order FIR filters are each filtering one of M multiplexed phases of the input stream, and the M outputs are being summed. This viewpoint offers a different implementation that might be advantageous in a multi-processor architecture. In other words, the input stream is demultiplexed and sent through a bank of M filters whose outputs are summed. When implemented that way, it is called a polyphase filter.

For completeness, we now mention that a possible, but unlikely, implementation of each phase is to replace the coefficients of the other phases with zeros in a copy of the h[•] array, process the original x[•] sequence at the input rate, and decimate the output by a factor of M. The equivalence of this inefficient method and the implementation described above is known as the *first Noble identity*.

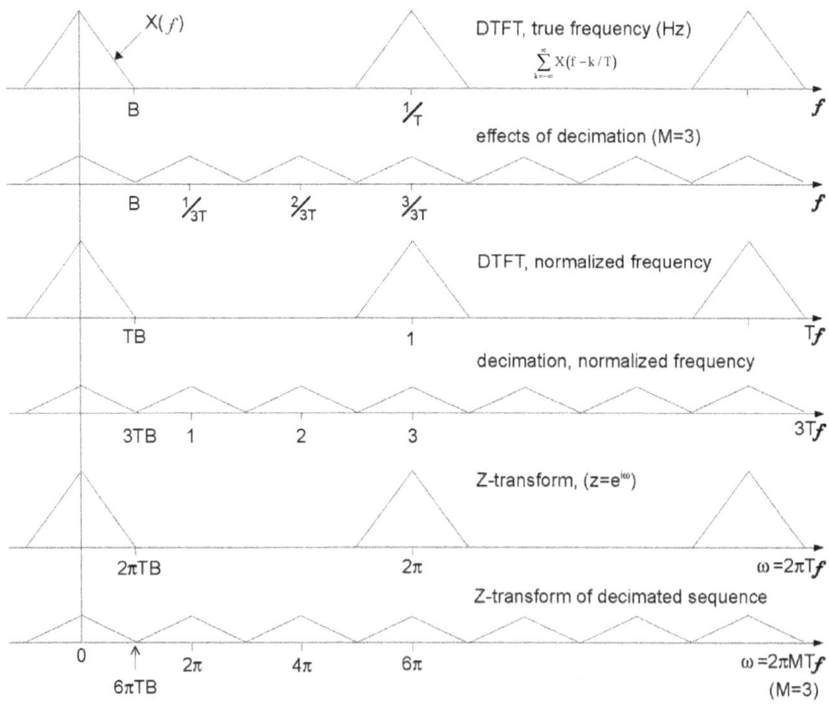

Figure 1: Spectral effects of decimation compared on 3 popular frequency scale conventions.

Anti-aliasing filter

The requirements of the anti-aliasing filter can be deduced from any of the 3 pairs of graphs in Fig. 1. Note that all 3 pairs are identical, except for the units of the abscissa variables. The upper graph of each pair is an example of the periodic frequency distribution of a sampled function, x(t), with Fourier transform, X(f). The lower graph is the new distribution that results when x(t) is sampled 3 times slower, or (equivalently) when the original sample sequence is decimated by a factor of M=3. In all 3 cases, the condition that ensures the copies of X(f) don't overlap each other is the same: $B < \frac{1}{M} \cdot \frac{1}{2T}$ where T is the interval between samples, 1/T is the sample-rate, and 1/2T is the Nyquist frequency. The anti-aliasing filter that can ensure the condition is met has a cutoff frequency less than $\frac{1}{M}$ times the Nyquist frequency.

The abscissa of the top pair of graphs represents the discrete-time Fourier transform (DTFT), which is a Fourier series representation of a periodic summation of X(f):

$$\underbrace{\sum_{n=-\infty}^{\infty} \overbrace{x(nT)}^{x[n]}\, e^{-i2\pi f nT}}_{\text{DTFT}} = \frac{1}{T}\sum_{k=-\infty}^{\infty} X(f - k/T).$$

When T has units of seconds, f has units of hertz. Replacing T with MT in the formulas above gives the DTFT of the decimated sequence, x[nM]:

$$\sum_{n=-\infty}^{\infty} x(n\cdot MT)\, e^{-i2\pi f n(MT)} = \frac{1}{MT}\sum_{k=-\infty}^{\infty} X\left(f - \frac{k}{(MT)}\right).$$

The periodic summation has been reduced in amplitude and periodicity by a factor of M, as depicted in the second graph of Fig. 1. Aliasing occurs when adjacent copies of X(f) overlap. The purpose of the anti-aliasing filter is to ensure that the reduced periodicity does not create overlap.

In the middle pair of graphs, the frequency variable, f has been replaced by normalized frequency, which creates a periodicity of 1 and a Nyquist frequency of ½. A common practice in filter design programs is to assume those values and request only the corresponding cutoff frequency in the same units. In other words, the cutoff frequency $B_{max} = \frac{1}{M} \cdot \frac{1}{2T}$, is normalized to $TB_{max} = \frac{1}{M} \cdot \frac{1}{2} = \frac{0.5}{M}$. The units of this quantity are (seconds/sample)×(cycles/second) = cycles/sample.

The bottom pair of graphs represent the Z-transforms of the original sequence and the decimated sequence, constrained to values of complex-variable, z, of the form $z = e^{i\omega}$ Then the transform of the x[n] sequence has the form of a Fourier series. By comparison with Eq.1, we deduce:

$$\sum_{n=-\infty}^{\infty} x[n]\, z^{-n} = \sum_{n=-\infty}^{\infty} x(nT)\, e^{-i\omega n} = \frac{1}{T}\sum_{k=-\infty}^{\infty} \underbrace{X\left(\frac{\omega}{2\pi T} - \frac{k}{T}\right)}_{X\left(\frac{\omega - 2\pi k}{2\pi T}\right)},$$

which is depicted by the fifth graph in Fig. 1. Similarly, the sixth graph depicts:

$$\sum_{n=-\infty}^{\infty} x[nM]\, z^{-n} = \sum_{n=-\infty}^{\infty} x(nMT)\, e^{-i\omega n} = \frac{1}{MT}\sum_{k=-\infty}^{\infty} \underbrace{X\left(\frac{\omega}{2\pi MT} - \frac{k}{MT}\right)}_{X\left(\frac{\omega - 2\pi k}{2\pi MT}\right)}.$$

By a Rational Factor

Let M/L denote the decimation factor, where: M, L ∈ Z; M > L.

- Interpolate by a factor of L
- Decimate by a factor of M

Interpolation requires a lowpass filter after increasing the data rate, and decimation requires a lowpass filter before decimation. Therefore, both operations can be accomplished by a single filter with the lower of the two cutoff frequencies. For the M > L case, the anti-aliasing filter cutoff, $\frac{0.5}{M}$ cycles per intermediate sample, is the lower frequency.

By an Irrational Factor

Techniques for decimation (and sample-rate conversion in general) by factor R ∈ R+ include polynomial interpolation and the Farrow structure.

REFERENCES

1. Martin H. Weik (1996). Communications Standard Dictionary. Springer. ISBN 0412083914.
2. Rao, R. Signals and Systems. Prentice-Hall Of India Pvt. Limited. ISBN 9788120338593.
3. C. E. Shannon, "Communication in the presence of noise", Proc. Institute of Radio Engineers, vol. 37, no.1, pp. 10–21, Jan. 1949. Reprint as classic paper in: Proc. IEEE, Vol. 86, No. 2, (Feb 1998)
4. Frequency Range of Human Hearing. The Physics Factbook.
5. Self, Douglas (2012). Audio Engineering Explained. Taylor & Francis US. pp. 200, 446. ISBN 0240812735.
6. Digital Pro Sound. Retrieved 8 January 2014.
7. Colletti, Justin (February 4, 2013). The Science of Sample Rates (When Higher Is Better—And When It Isn't). Trust Me I'm A Scientist. Retrieved February 6, 2013.
8. David Griesinger. Perception of mid frequency and high frequency intermodulation distortion in loudspeakers, and its relationship to high-definition audio. Archived from the original (Powerpoint presentation) on 2008-05-01.
9. AES5-2008: AES recommended practice for professional digital audio - Preferred sampling frequencies for applications employing pulse-code modulation, Audio Engineering Society, 2008, retrieved 2010-01-18
10. http://www.voipsupply.com/cisco-hd-voice
11. The restoration procedure - part 1. Restoring78s.co.uk. Archived from the original on 2009-09-14. Retrieved 2011-01-18. For most records a sample rate of 22050 in stereo is adequate. An exception is likely to be recordings made in the second half of the century, which may need a sample rate of 44100.
12. Zaxcom digital wireless transmitters. Zaxcom.com. Retrieved 2011-01-18.

13. MT-001: Taking the Mystery out of the Infamous Formula, SNR=6.02N + 1.76dB, and Why You Should Care (PDF).

14. Walt Kester (2003). Mixed-signal and DSP design techniques. Newnes. p. 20. ISBN 978-0-7506-7611-3. Retrieved 8 January 2014.

15. William Morris Hartmann (1997). Signals, Sound, and Sensation. Springer. ISBN 1563962837.

16. Allen Gersho and Robert M. Gray, Vector Quantization and Signal Compression, Springer, ISBN 978-0-7923-9181-4, 1991.

17. Hodgson, Jay (2010). Understanding Records, p.56. ISBN 978-1-4411-5607-5. Adapted from Franz, David (2004). Recording and Producing in the Home Studio, p.38-9. Berklee Press.

18. William Fleetwood Sheppard, "On the Calculation of the Most Probable Values of Frequency Constants for data arranged according to Equidistant Divisions of a Scale", Proceedings of the London Mathematical Society, Vol. 29, pp. 353–80, 1898.doi:10.1112/plms/s1-29.1.353

19. W. R. Bennett, Spectra of Quantized Signals, Bell System Technical Journal, Vol. 27, pp. 446–472, July 1948.

20. B. M. Oliver, J. R. Pierce, and Claude E. Shannon, The Philosophy of PCM, Proceedings of the IRE, Vol. 36, pp. 1324–1331, Nov. 1948. doi:10.1109/JRPROC.1948.231941

21. Seymour Stein and J. Jay Jones, Modern Communication Principles, McGraw–Hill, ISBN 978-0-07-061003-3, 1967 (p. 196).

22. Herbert Gish and John N. Pierce, "Asymptotically Efficient Quantizing", IEEE Transactions on Information Theory, Vol. IT-14, No. 5, pp. 676–683, Sept. 1968. doi:10.1109/TIT.1968.1054193

23. Robert M. Gray and David L. Neuhoff, Quantization, IEEE Transactions on Information Theory, Vol. IT-44, No. 6, pp. 2325–2383, Oct. 1998. doi:10.1109/18.720541

24. Allen Gersho, Quantization, IEEE Communications Society Magazine, pp. 16–28, Sept. 1977. doi:10.1109/MCOM.1977.1089500

25. Rabbani, Majid; Joshi, Rajan L.; Jones, Paul W. (2009). Section 1.2.3: Quantization, in Chapter 1: JPEG 2000 Core Coding System (Part 1). In Schelkens, Peter; Skodras, Athanassios; Ebrahimi, Touradj. The JPEG 2000 Suite. John Wiley & Sons. pp. 22–24. ISBN 978-0-470-72147-6.

26. Taubman, David S.; Marcellin, Michael W. (2002). Chapter 3: Quantization. JPEG2000: Image Compression Fundamentals, Standards and Practice. Kluwer Academic Publishers. p. 107. ISBN 0-7923-7519-X.

27. Gary J. Sullivan, "Efficient Scalar Quantization of Exponential and Laplacian Random Variables", IEEE Transactions on Information Theory, Vol. IT-42, No. 5, pp. 1365–1374, Sept. 1996. doi:10.1109/18.532878

28. Bernard Widrow, A study of rough amplitude quantization by means of Nyquist sampling theory, IRE Trans. Circuit Theory, Vol. CT-3, pp. 266–276, 1956. doi:10.1109/TCT.1956.1086334

29. Bernard Widrow, Statistical analysis of amplitude quantized sampled data systems, Trans. AIEE Pt. II: Appl. Ind., Vol. 79, pp. 555–568, Jan. 1961.

30. Daniel Marco and David L. Neuhoff, The Validity of the Additive Noise Model for Uniform Scalar Quantizers, IEEE Transactions on Information Theory, Vol. IT-51, No. 5, pp. 1739–1755, May 2005. doi:10.1109/TIT.2005.846397

31. Pohlman, Ken C. (1989). Principles of Digital Audio 2nd Edition. SAMS. p. 60.

32. Okelloto, Tom (2001). The Art of Digital Audio 3rd Edition. Focal Press. ISBN 0-240-51587-0.

33. Nariman Farvardin and James W. Modestino, Optimum Quantizer Performance for a Class of Non-Gaussian Memoryless Sources, IEEE Transactions on Information Theory, Vol. IT-30, No. 3, pp. 485–497, May 1982 (Section VI.C and Appendix B). doi:10.1109/TIT.1984.1056920

34. Toby Berger, "Optimum Quantizers and Permutation Codes, IEEE Transactions on Information Theory, Vol. IT-18, No. 6, pp. 759–765, Nov. 1972. doi:10.1109/TIT.1972.1054906

35. Toby Berger, "Minimum Entropy Quantizers and Permutation Codes, IEEE Transactions on Information Theory, Vol. IT-28, No. 2, pp. 149–157, Mar. 1982. doi:10.1109/TIT.1982.1056456

36. Stuart P. Lloyd, "Least Squares Quantization in PCM", IEEE Transactions on Information Theory, Vol. IT-28, pp. 129–137, No. 2, March 1982 doi:10.1109/TIT.1982.1056489 (work documented in a manuscript circulated for comments at Bell Laboratories with a department log date of 31 July 1957 and also presented at the 1957 meeting of the Institute of Mathematical Statistics, although not formally published until 1982).

37. Joel Max, "Quantizing for Minimum Distortion", IRE Transactions on Information Theory, Vol. IT-6, pp. 7–12, March 1960. doi:10.1109/TIT.1960.1057548

38. Philip A. Chou, Tom Lookabaugh, and Robert M. Gray, Entropy-Constrained Vector Quantization, IEEE Transactions on Acoustics, Speech, and Signal Processing, Vol. ASSP-37, No. 1, Jan. 1989. doi:10.1109/29.17498

CHAPTER 6

Data Compression

Chapter Contents

- **Introduction**
- **Lossless**
- **Lossy**
- **Theory**
- **Uses**
- **History**
- **Outlook And Currently Unused Potential**

Compression is useful because it reduces resources required to store and transmit data. Computational resources are consumed in the compression process and, usually, in the reversal of the process (decompression). Data compression is subject to a space–time complexity trade-off. For instance, a compression scheme for video may require expensive hardware for the video to be decompressed fast enough to be viewed as it is being decompressed, and the option to decompress the video in full before watching it may be inconvenient or require additional storage. The design of data compression schemes involves trade-offs among various factors, including the degree of compression, the amount of distortion introduced (when using lossy data compression), and the computational resources required to compress and decompress the data.

INTRODUCTION

In signal processing, data compression, source coding, or bit-rate reduction involves encoding information using fewer bits than the original representation. Compression can be either lossy or lossless. Lossless compression reduces bits by identifying and eliminating statistical redundancy. No information is lost in lossless compression. Lossy compression reduces bits by removing unnecessary or less important information. The process of reducing the size of a data file is referred to as data compression. In the context of data transmission, it is called source coding (encoding done at the source of the data before it is stored or transmitted) in opposition to channel coding. Compression is useful because it reduces resources required to store and transmit data. Computational resources are consumed in the compression process and, usually, in the reversal of the process (decompression). Data compression is subject to a space–time complexity trade-off. For instance, a compression scheme for video may require expensive hardware for the video to be decompressed fast enough to be viewed as it is being decompressed, and the option to decompress the video in full before watching it may be inconvenient or require additional storage. The design of data compression schemes involves trade-offs among various factors, including the degree of compression, the amount of distortion introduced (when using lossy data compression), and the computational resources required to compress and decompress the data.

LOSSLESS

Lossless data compression algorithms usually exploit statistical redundancy to represent data without losing any information, so that the process is reversible. Lossless compression is possible because most real-world data exhibits statistical redundancy. For example, an image may have areas of color that do not change over several pixels; instead of coding red pixel, red pixel, ... the data may be encoded as 279 red pixels. This is a basic example of run-length encoding; there are many schemes to reduce file size by eliminating redundancy.

The Lempel–Ziv (LZ) compression methods are among the most popular algorithms for lossless storage. DEFLATE is a variation on LZ optimized for decompression speed and compression ratio, but compression can be slow. DEFLATE is used in PKZIP, Gzip, and PNG. LZW (Lempel–Ziv–Welch) is used in GIF images. Also noteworthy is the LZR (Lempel-Ziv–Renau) algorithm, which serves as the basis for the Zip method. LZ methods use a table-based compression model where table entries are substituted for repeated strings of data. For most LZ methods, this table is generated dynamically from earlier data in the input. The table itself is often Huffman encoded (e.g. SHRI, LZX). Current LZ-based coding schemes that perform well are Brotli and LZX. LZX is used in Microsoft's CAB format. The best modern lossless compressors use probabilistic models, such as prediction by partial matching. The Burrows–Wheeler transform can also be viewed as an indirect form of statistical modelling.

The class of grammar-based codes are gaining popularity because they can compress highly repetitive input extremely effectively, for instance, a biological data collection of the same or closely related species, a huge versioned document collection, internet archival, etc. The basic task of grammar-based codes is constructing a context-free grammar deriving a single string. Sequitur and Re-Pair are practical grammar compression algorithms for which software is publicly available. In a further refinement of the direct use of probabilistic modelling, statistical estimates can be coupled to an algorithm called arithmetic coding. Arithmetic coding is a more modern coding technique that uses the mathematical calculations of a finite-state machine to produce a string of encoded bits from a series of input data symbols. It can achieve superior compression to other techniques such as the better-known Huffman algorithm. It uses an internal memory state to avoid the need to perform a one-to-one mapping of individual input symbols to distinct representations that use an integer number of bits, and it clears out the internal memory only after encoding the entire string of data symbols. Arithmetic coding applies especially well to adaptive data compression tasks where the statistics vary and are context-dependent, as it can be easily coupled with an adaptive model of the probability distribution of the input data. An early example of the use of arithmetic coding was its use as an optional (but not widely used) feature of the JPEG image coding standard. It has since been applied in various other designs including H.264/MPEG-4 AVC and HEVC for video coding.

LOSSY

Lossy data compression is the converse of lossless data compression. In these schemes, some loss of information is acceptable. Dropping nonessential detail from the data source can save storage space. Lossy data compression schemes are designed by research on how people perceive the data in question. For example, the human eye is more sensitive to subtle variations in luminance than it is to the variations in color. JPEG image compression works in part by rounding off nonessential bits of information. There is a corresponding trade-off between preserving information and reducing size. A number of popular compression formats exploit these perceptual differences, including those used in music files, images, and video.

Lossy image compression can be used in digital cameras, to increase storage capacities with minimal degradation of picture quality. Similarly, DVDs use the lossy MPEG-2 video coding format for video compression.

In lossy audio compression, methods of psychoacoustics are used to remove non-audible (or less audible) components of the audio signal. Compression of human speech is often performed with even more specialized techniques; speech coding, or voice coding, is sometimes distinguished as a separate discipline from audio compression. Different audio and speech compression standards are listed under audio coding formats. Voice compression is used in internet telephony, for example, audio compression is used for CD ripping and is decoded by the audio players.

THEORY

The theoretical background of compression is provided by information theory (which is closely related to algorithmic information theory) for lossless compression and rate–distortion theory for lossy compression. These areas of study were essentially forged by Claude Shannon, who published fundamental papers on the topic in the late 1940s and early 1950s. Coding theory is also related to this. The idea of data compression is also deeply connected with statistical inference.

Machine learning

There is a close connection between machine learning and compression: a system that predicts the posterior probabilities of a sequence given its entire history can be used for optimal data compression (by using arithmetic coding on the output distribution) while an optimal compressor can be used for prediction (by finding the symbol that compresses best, given the previous history). This equivalence has been used as a justification for using data compression as a benchmark for "general intelligence."

Data differencing

Data compression can be viewed as a special case of data differencing: Data differencing consists of producing a difference given a source and a target, with patching producing a target given a source and a difference, while data compression consists of producing a compressed file given a target, and decompression consists of producing a target given only a compressed file. Thus, one can consider data compression as data differencing with empty source data, the compressed file corresponding to a "difference from nothing." This is the same as considering absolute entropy (corresponding to data compression) as a special case of relative entropy (corresponding to data differencing) with no initial data.

When one wishes to emphasize the connection, one may use the term differential compression to refer to data differencing.

USES

Audio

Audio data compression, not to be confused with dynamic range compression, has the potential to reduce the transmission bandwidth and storage requirements of audio data. Audio compression algorithms are implemented in software as audio codecs. Lossy audio compression algorithms provide higher compression at the cost of fidelity and are used in numerous audio applications. These algorithms almost all rely on psychoacoustics to eliminate or reduce fidelity of less audible sounds, thereby reducing the space required to store or transmit them. In both lossy and lossless compression, information redundancy is reduced, using methods such as coding, pattern recognition, and linear prediction to reduce the amount of information used to represent the uncompressed data. The acceptable trade-off between loss of audio quality and transmission or storage size depends upon the application. For example, one 640MB compact disc (CD) holds approximately one hour of uncompressed high fidelity music, less than 2 hours of music compressed losslessly, or 7 hours of music compressed in the MP3 format at a medium bit rate. A digital sound recorder can typically store around 200 hours of clearly intelligible speech in 640MB.

Lossless audio compression produces a representation of digital data that decompress to an exact digital duplicate of the original audio stream, unlike playback from lossy compression techniques such as Vorbis and MP3. Compression ratios are around 50–60% of original size, which is similar to those for generic lossless data compression. Lossless compression is unable to attain high compression ratios due to the complexity of waveforms and the rapid changes in sound forms. Codecs like FLAC, Shorten, and TTA use linear prediction to estimate the spectrum of the signal. Many of these algorithms use convolution with the filter [-1 1] to slightly whiten or flatten the spectrum, thereby allowing traditional lossless compression to work more efficiently. The process is reversed upon decompression.

When audio files are to be processed, either by further compression or for editing, it is desirable to work from an unchanged original (uncompressed or losslessly compressed). Processing of a lossily compressed file for some purpose usually produces a final result inferior to the creation of the same compressed file from an uncompressed original. In addition to sound editing or mixing, lossless audio compression is often used for archival storage, or as master copies.

A number of lossless audio compression formats exist. Shorten was an early lossless format. Newer ones include Free Lossless Audio Codec (FLAC), Apple's Apple Lossless (ALAC), MPEG-4 ALS, Microsoft's Windows Media Audio 9 Lossless (WMA Lossless), Monkey's Audio, TTA, and WavPack. See list of lossless codecs for a complete listing.

Some audio formats feature a combination of a lossy format and a lossless correction; this allows stripping the correction to easily obtain a lossy file. Such formats include MPEG-4 SLS (Scalable to Lossless), WavPack, and OptimFROG DualStream.

Other formats are associated with a distinct system, such as:

- Direct Stream Transfer, used in Super Audio CD
- Meridian Lossless Packing, used in DVD-Audio, Dolby TrueHD, Blu-ray and HD DVD

Lossy audio compression

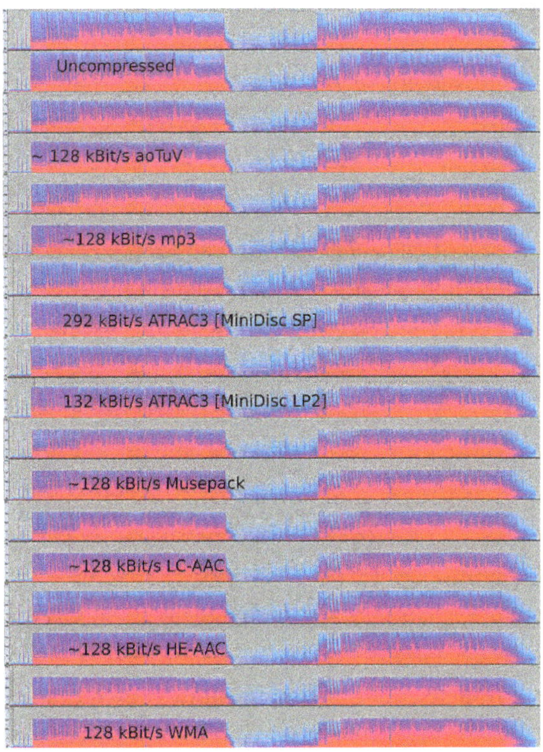

Comparison of acoustic spectrograms of a song in an uncompressed format and lossy formats. That the lossy spectrograms are different from the uncompressed one indicates that they are, in fact, lossy, but nothing can be assumed about the effect of the changes on perceived quality.

Lossy audio compression is used in a wide range of applications. In addition to the direct applications (MP3 players or computers), digitally compressed audio streams are used in most video DVDs, digital television, streaming media on the internet, satellite and cable radio, and increasingly in terrestrial radio broadcasts. Lossy compression typically achieves far greater compression than lossless compression (data of 5 percent to 20 percent of the original stream, rather than 50 percent to 60 percent), by discarding less-critical data.

The innovation of lossy audio compression was to use psychoacoustics to recognize that not all data in an audio stream can be perceived by the human auditory system. Most lossy compression reduces perceptual redundancy by first identifying perceptually irrelevant sounds, that is, sounds that are very hard to hear. Typical examples include high frequencies or sounds that occur at the same time as louder sounds. Those sounds are coded with decreased accuracy or not at all.

Due to the nature of lossy algorithms, audio quality suffers when a file is decompressed and recompressed (digital generation loss). This makes lossy compression unsuitable for storing the intermediate results in professional audio engineering applications, such as sound editing and multitrack recording. However, they are very popular with end users (particularly MP3) as a megabyte can store about a minute's worth of music at adequate quality.

Coding Methods

To determine what information in an audio signal is perceptually irrelevant, most lossy compression algorithms use transforms such as the modified discrete cosine transform (MDCT) to convert time domain sampled waveforms into a transform domain. Once transformed, typically into the frequency domain, component frequencies can be allocated bits according to how audible they are. Audibility of spectral components calculated using the absolute threshold of hearing and the principles of simultaneous masking—the phenomenon wherein a signal is masked by another signal separated by frequency—and, in some cases, temporal masking—where a signal is masked by another signal separated by time. Equal-loudness contours may also be used to weight the perceptual importance of components. Models of the human ear-brain combination incorporating such effects are often called psychoacoustic models.

Other types of lossy compressors, such as the linear predictive coding (LPC) used with speech, are source-based coders. These coders use a model of the sound's generator (such as the human vocal tract with LPC) to whiten the audio signal (i.e., flatten its spectrum) before quantization. LPC may be thought of as a basic perceptual coding technique: reconstruction of an audio signal using a linear predictor shapes the coder's quantization noise into the spectrum of the target signal, partially masking it.

Lossy formats are often used for the distribution of streaming audio or interactive applications (such as the coding of speech for digital transmission in cell phone networks). In such applications, the data must be decompressed as the data flows, rather than after the entire data stream has been transmitted. Not all audio codecs can be used for streaming applications, and for such applications a codec designed to stream data effectively will usually be chosen.

Latency results from the methods used to encode and decode the data. Some codecs will analyze a longer segment of the data to optimize efficiency, and then code it in a manner that requires a larger segment of data at one time to decode. (Often codecs create segments called a frame to create discrete data segments for encoding and decoding.) The inherent latency of the coding algorithm can be critical; for example, when there is a two-way transmission of data, such as with a telephone conversation, significant delays may seriously degrade the perceived quality.

In contrast to the speed of compression, which is proportional to the number of operations required by the algorithm, here latency refers to the number of samples that must be analysed before a block of audio is processed. In the minimum case, latency is zero samples (e.g., if the coder/decoder simply reduces the number of bits used to quantize the signal). Time domain algorithms such as LPC also often have low latencies, hence their popularity in speech coding for telephony. In algorithms such as MP3, however, a large number of samples have to be analyzed to implement a psychoacoustic model in the frequency domain, and latency is on the order of 23 ms (46 ms for two-way communication)).

Speech Encoding

Speech encoding is an important category of audio data compression. The perceptual models used to estimate what a human ear can hear are generally somewhat different from those used for music. The range of frequencies needed to convey the sounds of a human voice are normally far narrower than that needed for music, and the sound is normally less complex. As a result, speech can be encoded at high quality using a relatively low bit rate.

If the data to be compressed is analog (such as a voltage that varies with time), quantization is employed to digitize it into numbers (normally integers). This is referred to as analog-to-digital (A/D) conversion. If the integers generated by quantization are 8 bits each, then the entire range of the analog signal is divided into 256 intervals and all the signal values within an interval are quantized to the same number. If 16-bit integers are generated, then the range of the analog signal is divided into 65,536 intervals.

This relation illustrates the compromise between high resolution (a large number of analog intervals) and high compression (small integers generated). This application of quantization is used by several speech compression methods. This is accomplished, in general, by some combination of two approaches:

- Only encoding sounds that could be made by a single human voice.
- Throwing away more of the data in the signal—keeping just enough to reconstruct an intelligible voice rather than the full frequency range of human hearing.

Perhaps the earliest algorithms used in speech encoding (and audio data compression in general) were the A-law algorithm and the μ-law algorithm.

HISTORY

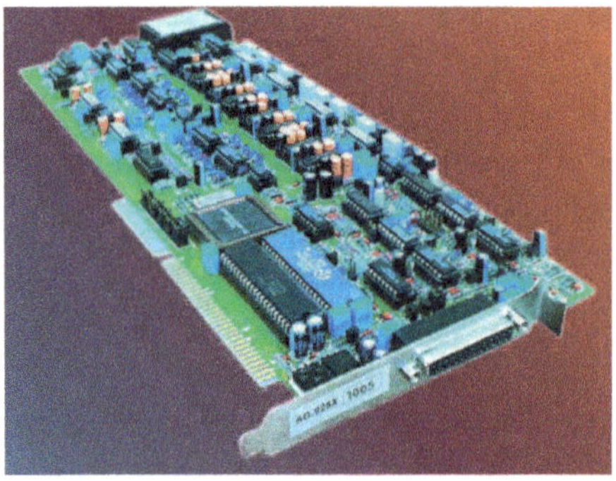

Solidyne 922: The world's first commercial audio bit compression card for PC, 1990

A literature compendium for a large variety of audio coding systems was published in the IEEE Journal on Selected Areas in Communications (JSAC), February 1988. While there were some papers from before that time, this collection documented an entire variety of finished, working audio coders, nearly all of them using perceptual (i.e. masking) techniques and some kind of frequency analysis and back-end noiseless coding.Several of these papers remarked on the difficulty of obtaining good, clean digital audio for research purposes. Most, if not all, of the authors in the JSAC edition were also active in the MPEG-1 Audio committee.

The world's first commercial broadcast automation audio compression system was developed by Oscar Bonello, an engineering professor at the University of Buenos Aires. In 1983, using the psychoacoustic principle of the masking of critical bands first published in 1967, he started developing a practical application based on the recently developed IBM PC computer, and the broadcast automation system was launched in 1987 under the name Audicom. Twenty years later, almost all the radio stations in the world were using similar technology manufactured by a number of companies.

Video

Video compression uses modern coding techniques to reduce redundancy in video data. Most video compression algorithms and codecs combine spatial image compression and temporal motion compensation. Video compression is a practical implementation of source coding in information theory. In practice, most video codecs also use audio compression techniques in parallel to compress the separate, but combined data streams as one package.

The majority of video compression algorithms use lossy compression. Uncompressed video requires a very high data rate. Although lossless video compression codecs perform at a compression factor of 5-12, a typical MPEG-4 lossy compression video has a compression factor between 20 and 200. As in all lossy compression, there is a trade-off between video quality, cost of processing the compression and decompression, and system requirements. Highly compressed video may present visible or distracting artifacts.

Some video compression schemes typically operate on square-shaped groups of neighboring pixels, often called macroblocks. These pixel groups or blocks of pixels are compared from one frame to the next, and the video compression codec sends only the differences within those blocks. In areas of video with more motion, the compression must encode more data to keep up with the larger number of pixels that are changing. Commonly during explosions, flames, flocks of animals, and in some panning shots, the high-frequency detail leads to quality decreases or to increases in the variable bitrate.

Encoding Theory

Video data may be represented as a series of still image frames. The sequence of frames contains spatial and temporal redundancy that video compression algorithms attempt to eliminate or code in a smaller size. Similarities can be encoded by only storing differences between frames, or by using perceptual features of human vision. For example, small differences in color are more difficult to perceive than are changes in brightness. Compression algorithms can average a color across these similar areas to reduce space, in a manner similar to those used in JPEG image compression. Some of these methods are inherently lossy while others may preserve all relevant information from the original, uncompressed video.

One of the most powerful techniques for compressing video is interframe compression. Interframe compression uses one or more earlier or later frames in a sequence to compress the current frame, while intraframe compression uses only the current frame, effectively being image compression.

The most powerful used method works by comparing each frame in the video with the previous one. If the frame contains areas where nothing has moved, the system simply issues a short command that copies that part of the previous frame, bit-for-bit, into the next one. If sections of the frame move in a simple manner, the compressor emits a (slightly longer) command that tells the decompressor to shift, rotate, lighten, or darken the copy. This longer command still remains much shorter than intraframe compression. Interframe compression works well for programs that will simply be played back by the viewer, but can cause problems if the video sequence needs to be edited.

Because interframe compression copies data from one frame to another, if the original frame is simply cut out (or lost in transmission), the following frames cannot be reconstructed properly. Some video formats, such as DV, compress each frame independently using intraframe compression. Making 'cuts' in intraframe-compressed video is almost as easy as editing uncompressed video: one finds the beginning and ending of each frame, and simply copies bit-for-bit each frame that one wants to keep, and discards the frames one doesn't want. Another difference between intraframe and interframe compression is that, with intraframe systems, each frame uses a similar amount of data. In most interframe systems, certain frames (such as I frames in MPEG-2) aren't allowed to copy data from other frames, so they require much more data than other frames nearby.

It is possible to build a computer-based video editor that spots problems caused when I frames are edited out while other frames need them. This has allowed newer formats like HDV to be used for editing. However, this process demands a lot more computing power than editing intraframe compressed video with the same picture quality.

Today, nearly all commonly used video compression methods (e.g., those in standards approved by the ITU-T or ISO) apply a discrete cosine transform (DCT) for spatial redundancy reduction. The DCT that is widely used in this regard was introduced by N. Ahmed, T. Natarajan and K. R. Rao in 1974. Other methods, such as fractal compression, matching pursuit and the use of a discrete wavelet transform (DWT) have been the subject of some research, but are typically not used in practical products (except for the use of wavelet coding as still-image coders without motion compensation). Interest in fractal compression seems to be waning, due to recent theoretical analysis showing a comparative lack of effectiveness of such methods.

Timeline

The following table is a partial history of international video compression standards.

History of video compression standards			
Year	**Standard**	**Publisher**	**Popular implementations**
1984	H.120	ITU-T	
1988	H.261	ITU-T	Videoconferencing, videotelephony
1993	MPEG-1 Part 2	ISO, IEC	Video-CD
1995	H.262/MPEG-2 Part 2	ISO, IEC, ITU-T	DVD Video, Blu-ray, Digital Video Broadcasting, SVCD
1996	H.263	ITU-T	Videoconferencing, videotelephony, video on mobile phones (3GP)
1999	MPEG-4 Part 2	ISO, IEC	Video on Internet (DivX, Xvid ...)
2003	H.264/MPEG-4 AVC	Sony, Panasonic, Samsung, ISO, IEC, ITU-T	Blu-ray, HD DVD, Digital Video Broadcasting, iPod Video, Apple TV, videoconferencing
2009	VC-2 (Dirac)	SMPTE	Video on Internet, HDTV broadcast, UHDTV
2013	H.265	ISO, IEC, ITU-T	

Genetics

Genetics compression algorithms are the latest generation of lossless algorithms that compress data (typically sequences of nucleotides) using both conventional compression algorithms and genetic algorithms adapted to the specific datatype. In 2012, a team of scientists from Johns Hopkins University published a genetic compression algorithm that does not use a reference genome for compression. HAPZIPPER was tailored for HapMap data and achieves over 20-fold compression (95% reduction in file size), providing 2- to 4-fold better compression and in much faster time than the leading general-purpose compression utilities. For this, Chanda, Elhaik, and Bader introduced MAF based encoding (MAFE), which reduces the heterogeneity of the dataset by sorting SNPs by their minor allele frequency, thus homogenizing the dataset. Other algorithms in 2009 and 2013 (DNAZip and GenomeZip) have compression ratios of up to 1200-fold—allowing 6 billion basepair diploid human genomes to be stored in 2.5 megabytes (relative to a reference genome or averaged over many genomes).

Emulation

In order to emulate CD-based consoles such as the PlayStation 2, data compression is desirable to reduce huge amounts of disk space used by ISOs. For example, Final Fantasy XII (USA) is normally 2.9 gigabytes. With proper compression, it is reduced to around 90% of that size.

OUTLOOK AND CURRENTLY UNUSED POTENTIAL

It is estimated that the total amount of data that is stored on the world's storage devices could be further compressed with existing compression algorithms by a remaining average factor of 4.5:1. It is estimated that the combined technological capacity of the world to store information provides 1,300 exabytes of hardware digits in 2007, but when the corresponding content is optimally compressed, this only represents 295 exabytes of Shannon information.

REFERENCES

1. Wade, Graham (1994). Signal coding and processing (2 ed.). Cambridge University Press. p. 34. ISBN 978-0-521-42336-6. Retrieved 2011-12-22. The broad objective of source coding is to exploit or remove 'inefficient' redundancy in the PCM source and thereby achieve a reduction in the overall source rate R.
2. Mahdi, O.A.; Mohammed, M.A.; Mohamed, A.J. (November 2012). "Implementing a Novel Approach an Convert Audio Compression to Text Coding via Hybrid Technique" (PDF). International Journal of Computer Science Issues. **9** *(6, No. 3): 53–59. Retrieved 6 March 2013.*
3. Pujar, J.H.; Kadlaskar, L.M. (May 2010). A New Lossless Method of Image Compression and Decompression Using Huffman Coding Techniques (PDF). Journal of Theoretical and Applied Information Technology. **15** *(1): 18–23.*
4. Salomon, David (2008). A Concise Introduction to Data Compression. Berlin: Springer. ISBN 9781848000728.
5. S. Mittal and J. Vetter, "A Survey Of Architectural Approaches for Data Compression in Cache and Main Memory Systems, IEEE Transactions on Parallel and Distributed Systems, 2015.
6. Tank, M.K. (2011). Implementation of Limpel-Ziv algorithm for lossless compression using VHDL. Thinkquest 2010: Proceedings of the First International Conference on Contours of Computing Technology. Berlin: Springer. pp. 275–283.
7. Navqi, Saud; Naqvi, R.; Riaz, R.A.; Siddiqui, F. (April 2011). Optimized RTL design and implementation of LZW algorithm for high bandwidth applications (PDF). Electrical Review. **2011** *(4): 279–285.*
8. Mahmud, Salauddin (March 2012). An Improved Data Compression Method for General Data(PDF). International Journal of Scientific & Engineering Research. **3** *(3): 2.*

Retrieved 6 March 2013.

9. Arcangel, Cory. On Compression (PDF). Retrieved 6 March 2013.
10. Marak, Laszlo. On image compression (PDF). University of Marne la Vallee. Retrieved 6 March 2013.
11. Mahoney, Matt. Rationale for a Large Text Compression Benchmark. Florida Institute of Technology. Retrieved 5 March 2013.
12. Shmilovici A.; Kahiri Y.; Ben-Gal I.; Hauser S. Measuring the Efficiency of the Intraday Forex Market with a Universal Data Compression Algorithm (PDF). Computational Economics, Vol. 33 (2), 131-154., 2009.
13. I. Ben-Gal. On the Use of Data Compression Measures to Analyze Robust Designs (PDF). IEEE Trans. on Reliability, Vol. 54, no. 3, 381-388, 2008.
14. Korn, D.; et al. RFC 3284: The VCDIFF Generic Differencing and Compression Data Format. Internet Engineering Task Force. Retrieved 5 March 2013.
15. Korn, D.G.; Vo, K.P. (1995), B. Krishnamurthy, ed., Vdelta: Differencing and Compression, Practical Reusable Unix Software, New York: John Wiley & Sons, Inc.
16. The Olympus WS-120 digital speech recorder, according to its manual, can store about 178 hours of speech-quality audio in .WMA format in 500MB of flash memory.
17. Coalson, Josh. FLAC Comparison. Retrieved 6 March 2013.
18. Jaiswal, R.C. (2009). Audio-Video Engineering. Pune, Maharashtra: Nirali Prakashan. p. 3.41. ISBN 9788190639675.
19. Faxin Yu; Hao Luo; Zheming Lu (2010). Three-Dimensional Model Analysis and Processing. Berlin: Springer. p. 47. ISBN 9783642126512.
20. File Compression Possibilities. A Brief guide to compress a file in 6 different ways.
21. Summary of some of Solidyne's contributions to Broadcast Engineering. Brief History of Solidyne. Buenos Aires: Solidyne. Archived from the original on 8 March 2013. Retrieved 6 March 2013.
22. Zwicker, Eberhard; et al. (1967). The Ear As A Communication Receiver. Melville, NY: Acoustical Society of America.
23. Video Coding. Center for Signal and Information Processing Research. Georgia Institute of Technology. Retrieved 6 March 2013.
24. Graphics & Media Lab Video Group (2007). Lossless Video Codecs Comparison (PDF). Moscow State University.
25. Lane, Tom. JPEG Image Compression FAQ, Part 1. Internet FAQ Archives. Independent JPEG Group. Retrieved 6 March 2013.
26. Faxin Yu; Hao Luo; Zheming Lu (2010). Three-Dimensional Model Analysis and Processing. Berlin: Springer. p. 47. ISBN 9783642126512.
27. Bhojani, D.R. 4.1 Video Compression (PDF). Hypothesis. Retrieved 6 March 2013.
28. Ahmed, N.; Natarajan, T.; Rao, K.R. (January 1974). Discrete Cosine Transform. IEEE Transactions on Computers. **C–23** *(1): 90–93. doi:10.1109/T-C.1974.223784.*
29. Chanda P, Bader JS, Elhaik E; Elhaik; Bader (27 Jul 2012). HapZipper: sharing HapMap populations just got easier (PDF). Nucleic Acids Research. **40** *(20): e159.*

*doi:10.1093/*nar/gks709. PMC 3488212. PMID 22844100.

30. Christley S, Lu Y, Li C, Xie X; Lu; Li; Xie (Jan 15, 2009). Human genomes as email attachments. Bioinformatics. **25** *(2): 274–5. doi:10.1093/bioinformatics/btn582. PMID 18996942.*

31. Pavlichin DS, Weissman T, Yona G; Weissman; Yona (September 2013). The human genome contracts again. Bioinformatics. **29** *(17): 2199–202. doi:10.1093/bioinformatics/btt362. PMID 23793748.*

32. PCSX2 team (January 8, 2016). 1.4.0 released! Year end report. Retrieved April 27, 2016.

33. Hilbert, Martin; López, Priscila (1 April 2011). The World's Technological Capacity to Store, Communicate, and Compute Information. Science. **332** *(6025): 60–65. Bibcode*:2011Sci...332...60H. doi:10.1126/science.1200970. PMID 21310967. Retrieved 6 March 2013.

CHAPTER 7

Kalman Filtering

Chapter Contents

Kalman filtering, also known as linear quadratic estimation (LQE), is an algorithm that uses a series of measurements observed over time, containing statistical noise and other inaccuracies, and produces estimates of unknown variables that tend to be more precise than those based on a single measurement alone, by using Bayesian inference and estimating a joint probability distribution over the variables for each timeframe. The filter is named after Rudolf E. Kálmán, one of the primary developers of its theory.

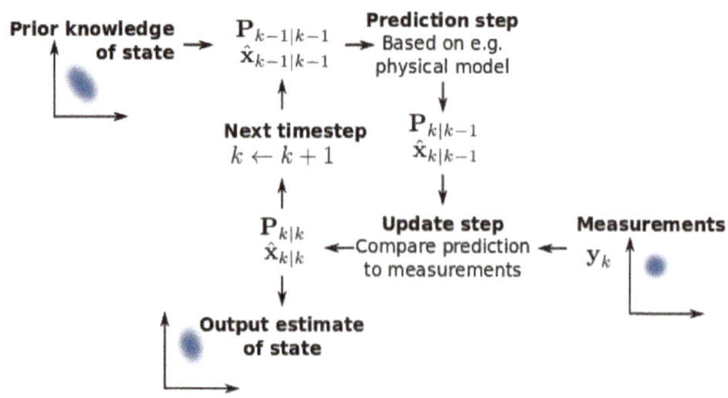

The Kalman filter keeps track of the estimated state of the system and the variance or uncertainty of the estimate. The estimate is updated using a state transition model and measurements. $\hat{x}_{k|k-1}$ denotes the estimate of the system's state at time step k before the k-th measurement y_k has been taken into account; $P_{k|k-1}$ is the corresponding uncertainty.

INTRODUCTION

The Kalman filter has numerous applications in technology. A common application is for guidance, navigation, and control of vehicles, particularly aircraft and spacecraft. Furthermore, the Kalman filter is a widely applied concept in time series analysis used in fields such as signal processing and econometrics. Kalman filters also are one of the main topics in the field of robotic motion planning and control, and they are sometimes included in trajectory optimization. The Kalman filter also works for modeling the central nervous system's control of movement. Due to the time delay between issuing motor commands and receiving sensory feedback, usage of the Kalman filter supports the realistic model for making estimates of the current state of the motor system and issuing updated commands.

The algorithm works in a two-step process. In the prediction step, the Kalman filter produces estimates of the current state variables, along with their uncertainties. Once the outcome of the next measurement (necessarily corrupted with some amount of error, including random noise) is observed, these estimates are updated using a weighted average, with more weight being given to estimates with higher certainty. The algorithm is recursive. It can run in real time, using only the present input measurements and the previously calculated state and its uncertainty matrix; no additional past information is required.

The Kalman filter does not make any assumption that the errors are Gaussian. However, the filter yields the exact conditional probability estimate in the special case that all errors are Gaussian-distributed.

Extensions and generalizations to the method have also been developed, such as the extended Kalman filter and the unscented Kalman filter which work on nonlinear systems. The underlying model is a Bayesian model similar to a hidden Markov model but where the state space of the latent variables is continuous and where all latent and observed variables have Gaussian distributions.

NAMING AND HISTORICAL DEVELOPMENT

The filter is named after Hungarian émigré Rudolf E. Kálmán, although Thorvald Nicolai Thiele and Peter Swerling developed a similar algorithm earlier. Richard S. Bucy of the University of Southern California contributed to the theory, leading to it often being called the Kalman–Bucy filter. Stanley F. Schmidt is generally credited with developing the first implementation of a Kalman filter. He realized that the filter could be divided into two distinct parts, with one part for time periods between sensor outputs and another part for incorporating measurements. It was during a visit by Kálmán to the NASA Ames Research Center that Schmidt saw the applicability of Kálmán's ideas to the nonlinear problem of trajectory estimation for the Apollo program leading to its incorporation in the Apollo navigation computer. This Kalman filter was first described and partially developed in technical papers by Swerling (1958), Kalman (1960) and Kalman and Bucy (1961).

Kalman filters have been vital in the implementation of the navigation systems of U.S. Navy nuclear ballistic missile submarines, and in the guidance and navigation systems of cruise missiles such as the U.S. Navy's Tomahawk missile and the U.S. Air Force's Air Launched Cruise Missile. It is also used in the guidance and navigation systems of reusable launch vehicles and the attitude control and navigation systems of spacecraft which dock at the International Space Station.

This digital filter is sometimes called the Stratonovich–Kalman–Bucy filter because it is a special case of a more general, non-linear filter developed somewhat earlier by the Soviet mathematician Ruslan Stratonovich. In fact, some of the special case linear filter's equations appeared in these papers by Stratonovich that were published before summer 1960, when Kalman met with Stratonovich during a conference in Moscow.

OVERVIEW OF THE CALCULATION

The Kalman filter uses a system's dynamics model (e.g., physical laws of motion), known control inputs to that system, and multiple sequential measurements (such as from sensors) to form an estimate of the system's varying quantities (its state) that is better than the estimate obtained by using only one measurement alone. As such, it is a common sensor fusion and data fusion algorithm.

Noisy sensor data, approximations in the equations that describe the system evolution, and external factors that are not accounted for all place limits on how well it is possible to determine the system's state. The Kalman filter deals effectively with the uncertainty due to noisy sensor data and to some extent also with random external factors. The Kalman filter produces an estimate of the state of the system as an average of the system's predicted state and of the new measurement using a weighted average. The purpose of the weights is that values with better (i.e., smaller) estimated uncertainty are trusted more. The weights are calculated from the covariance, a measure of the estimated uncertainty of the prediction of the system's state. The result of the weighted average is a new state estimate that lies between the predicted and measured state, and has a better estimated uncertainty than either alone. This process is repeated at every time step, with the new estimate and its covariance informing the prediction used in the following iteration. This means that the Kalman filter works recursively and requires only the last "best guess", rather than the entire history, of a system's state to calculate a new state.

The relative certainty of the measurements and current state estimate is an important consideration, and it is common to discuss the response of the filter in terms of the Kalman filter's *gain*. The Kalman gain is the relative weight given to the measurements and current state estimate, and can be "tuned" to achieve particular performance. With a high gain, the filter places more weight on the most recent measurements, and thus follows them more responsively. With a low gain, the filter follows the model predictions more closely. At the extremes, a high gain close to one will result in a more jumpy estimated trajectory, while low gain close to zero will smooth out noise but decrease the responsiveness.

When performing the actual calculations for the filter (as discussed below), the state estimate and covariances are coded into matrices to handle the multiple dimensions involved in a single set of calculations. This allows for a representation of linear relationships between different state variables (such as position, velocity, and acceleration) in any of the transition models or covariances.

EXAMPLE APPLICATION

As an example application, consider the problem of determining the precise location of a truck. The truck can be equipped with a GPS unit that provides an estimate of the position within a few meters. The GPS estimate is likely to be noisy; readings 'jump around' rapidly, though remaining within a few meters of the real position. In addition, since the truck is expected to follow the laws of physics, its position can also be estimated by integrating its velocity over time, determined by keeping track of wheel revolutions and the angle of the steering wheel. This is a technique known as dead reckoning. Typically, the dead reckoning will provide a very smooth estimate of the truck's position, but it will drift over time as small errors accumulate.

In this example, the Kalman filter can be thought of as operating in two distinct phases: predict and update. In the prediction phase, the truck's old position will be modified according to the physical laws of motion (the dynamic or "state transition" model). Not only will a new position estimate be calculated, but a new covariance will be calculated as well. Perhaps the covariance is proportional to the speed of the truck because we are more uncertain about the accuracy of the dead reckoning position estimate at high speeds but very certain about the position estimate when moving slowly. Next, in the update phase, a measurement of the truck's position is taken from the GPS unit. Along with this measurement comes some amount of uncertainty, and its covariance relative to that of the prediction from the previous phase determines how much the new measurement will affect the updated prediction. Ideally, as the dead reckoning estimates tend to drift away from the real position, the GPS measurement should pull the position estimate back towards the real position but not disturb it to the point of becoming rapidly jumping and noisy.

Technical description and context

The Kalman filter is an efficient recursive filter that estimates the internal state of a linear dynamic system from a series of noisy measurements. It is used in a wide range of engineering and econometric applications from radar and computer vision to estimation of structural macroeconomic models,and is an important topic in control theory and control systems engineering. Together with the linear-quadratic regulator (LQR), the Kalman filter solves the linear-quadratic-Gaussian control problem (LQG). The Kalman filter, the linear-quadratic regulator and the linear-quadratic-Gaussian controller are solutions to what arguably are the most fundamental problems in control theory.

In most applications, the internal state is much larger (more degrees of freedom) than the few observable parameters which are measured. However, by combining a series of measurements, the Kalman filter can estimate the entire internal state.

In Dempster–Shafer theory, each state equation or observation is considered a special case of a linear belief function and the Kalman filter is a special case of combining linear belief functions on a join-tree or Markov tree. Additional approaches include belief filters which use Bayes or evidential updates to the state equations.

A wide variety of Kalman filters have now been developed, from Kalman's original formulation, now called the simple Kalman filter, the Kalman–Bucy filter, Schmidt's extended filter, the information filter, and a variety of square-root filters that were developed by Bierman, Thornton and many others. Perhaps the most commonly used type of very simple Kalman filter is the phase-locked loop, which is now ubiquitous in radios, especially frequency modulation (FM) radios, television sets, satellite communications receivers, outer space communications systems, and nearly any other electronic communications equipment.

UNDERLYING DYNAMICAL SYSTEM MODEL

The Kalman filters are based on linear dynamical systems discretized in the time domain. They are modelled on a Markov chain built on linear operators perturbed by errors that may include Gaussiannoise. The state of the system is represented as a vector of real numbers. At each discrete time increment, a linear operator is applied to the state to generate the new state, with some noise mixed in, and optionally some information from the controls on the system if they are known. Then, another linear operator mixed with more noise generates the observed outputs from the true ("hidden") state. The Kalman filter may be regarded as analogous to the hidden Markov model, with the key difference that the hidden state variables take values in a continuous space (as opposed to a discrete state space as in the hidden Markov model). There is a strong duality between the equations of the Kalman Filter and those of the hidden Markov model. A review of this and other models is given in Roweis and Ghahramani (1999), and Hamilton (1994), Chapter 13.

In order to use the Kalman filter to estimate the internal state of a process given only a sequence of noisy observations, one must model the process in accordance with the framework of the Kalman filter. This means specifying the following matrices: $\mathbf{F}_k$, the state-transition model; $\mathbf{H}_k$, the observation model; $\mathbf{Q}_k$, the covariance of the process noise; $\mathbf{R}_k$, the covariance of the observation noise; and sometimes $\mathbf{B}_k$, the control-input model, for each time-step, k, as described below.

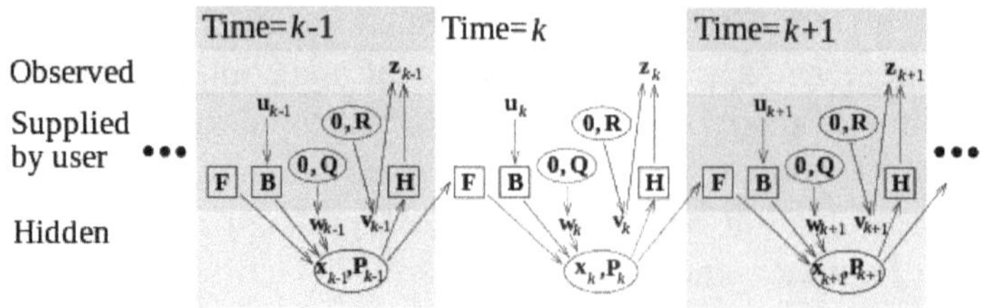

Model underlying the Kalman filter. Squares represent matrices. Ellipses represent multivariate normal distributions (with the mean and covariance matrix enclosed). Unenclosed values are vectors. In the simple case, the various matrices are constant with time, and thus the subscripts are dropped, but the Kalman filter allows any of them to change each time step.

The Kalman filter model assumes the true state at time k is evolved from the state at (k − 1) according to

$$\mathbf{x}_k = \mathbf{F}_k\mathbf{x}_{k-1} + \mathbf{B}_k\mathbf{u}_k + \mathbf{w}_k$$

where

- $\mathbf{F}_k$ is the state transition model which is applied to the previous state $\mathbf{x}_{k-1}$;
- $\mathbf{B}_k$ is the control-input model which is applied to the control vector $\mathbf{u}_k$;
- $\mathbf{w}_k$ is the process noise which is assumed to be drawn from a zero mean multivariate normal distribution with covariance $\mathbf{Q}_k$.

$$\mathbf{w}_k \sim \mathcal{N}(0, \mathbf{Q}_k)$$

At time k an observation (or measurement) zk of the true state xk is made according to

$$\mathbf{z}_k = \mathbf{H}_k\mathbf{x}_k + \mathbf{v}_k$$

where Hk is the observation model which maps the true state space into the observed space and vk is the observation noise which is assumed to be zero mean Gaussian white noise with covariance Rk.

$$\mathbf{v}_k \sim \mathcal{N}(0, \mathbf{R}_k)$$

The initial state, and the noise vectors at each step {x0, w1, ..., wk, v1 ... vk} are all assumed to be mutually independent.

Many real dynamical systems do not exactly fit this model. In fact, unmodelled dynamics can seriously degrade the filter performance, even when it was supposed to work with unknown stochastic signals as inputs. The reason for this is that the effect of unmodelled dynamics depends on the input, and, therefore, can bring the estimation algorithm to instability (it diverges). On the other hand, independent white noise signals will not make the algorithm diverge. The problem of distinguishing between measurement noise and unmodelled dynamics is a difficult one and is treated in control theory under the framework of robust control.

Details

The Kalman filter is a recursive estimator. This means that only the estimated state from the previous time step and the current measurement are needed to compute the estimate for the current state. In contrast to batch estimation techniques, no history of observations and/or estimates is required. In what follows, the notation $\hat{\mathbf{x}}_{n|m}$ represents the estimate of x at time n given observations up to and including at time m ≤ n.

The state of the filter is represented by two variables:

- $\hat{\mathbf{x}}_{k|k}$, the *a posteriori* state estimate at time *k* given observations up to and including at time *k*;
- $\mathbf{P}_{k|k}$, the *a posteriori* error covariance matrix (a measure of the estimated accuracy of the state estimate).

The Kalman filter can be written as a single equation, however it is most often conceptualized as two distinct phases: Predict and Update. The predict phase uses the state estimate from the previous timestep to produce an estimate of the state at the current timestep. This predicted state estimate is also known as the a priori state estimate because, although it is an estimate of the state at the current timestep, it does not include observation information from the current timestep. In the update phase, the current a priori prediction is combined with current observation information to refine the state estimate. This improved estimate is termed the a posteriori state estimate.

Typically, the two phases alternate, with the prediction advancing the state until the next scheduled observation, and the update incorporating the observation. However, this is not necessary; if an observation is unavailable for some reason, the update may be skipped and multiple prediction steps performed. Likewise, if multiple independent observations are available at the same time, multiple update steps may be performed (typically with different observation matrices $\mathbf{H}_k$).

Predicted (*a priori*) state estimate	$\hat{\mathbf{x}}_{k|k-1} = \mathbf{F}_k \hat{\mathbf{x}}_{k-1|k-1} + \mathbf{B}_k \mathbf{u}_k$
Predicted (*a priori*) estimate covariance	$\mathbf{P}_{k|k-1} = \mathbf{F}_k \mathbf{P}_{k-1|k-1} \mathbf{F}_k^{\mathrm{T}} + \mathbf{Q}_k$

Update

Innovation or measurement residual	$\tilde{\mathbf{y}}_k = \mathbf{z}_k - \mathbf{H}_k \hat{\mathbf{x}}_{k|k-1}$
Innovation (or residual) covariance	$\mathbf{S}_k = \mathbf{H}_k \mathbf{P}_{k|k-1} \mathbf{H}_k^T + \mathbf{R}_k$
Optimal Kalman gain	$\mathbf{K}_k = \mathbf{P}_{k|k-1} \mathbf{H}_k^T \mathbf{S}_k^{-1}$
Updated (*a posteriori*) state estimate	$\hat{\mathbf{x}}_{k|k} = \hat{\mathbf{x}}_{k|k-1} + \mathbf{K}_k \tilde{\mathbf{y}}_k$
Updated (*a posteriori*) estimate covariance	$\mathbf{P}_{k|k} = (I - \mathbf{K}_k \mathbf{H}_k) \mathbf{P}_{k|k-1}$

The formula for the updated estimate covariance above is only valid for the optimal Kalman gain. Usage of other gain values requires a more complex formula found in the derivations section.

Invariants

If the model is accurate, and the values for $\hat{\mathbf{x}}_{0|0}$ and $\mathbf{P}_{0|0}$ accurately reflect the distribution of the initial state values, then the following invariants are preserved:

- $\mathbf{E}[\mathbf{x}_k - \hat{\mathbf{x}}_{k|k}] = \mathbf{E}[\mathbf{x}_k - \hat{\mathbf{x}}_{k|k-1}] = 0$
- $\mathbf{E}[\tilde{\mathbf{y}}_k] = 0$

where $\mathrm{E}[\xi]$ is the expected value of ξ. That is, all estimates have a mean error of zero.

Also:

- $\mathbf{P}_{k|k} = \mathrm{cov}(\mathbf{x}_k - \hat{\mathbf{x}}_{k|k})$
- $\mathbf{P}_{k|k-1} = \mathrm{cov}(\mathbf{x}_k - \hat{\mathbf{x}}_{k|k-1})$
- $\mathbf{S}_k = \mathrm{cov}(\tilde{\mathbf{y}}_k)$

so covariance matrices accurately reflect the covariance of estimates.

Estimation of the Noise Covariances Q_k and R_k

Practical implementation of the Kalman Filter is often difficult due to the difficulty of getting a good estimate of the noise covariance matrices Qk and Rk. Extensive research has been done in this field to estimate these covariances from data. One of the more promising and practical approaches to do this is the autocovariance least-squares (ALS) technique that uses the time-lagged autocovariances of routine operating data to estimate the covariances. The GNU Octave and Matlab code used to calculate the noise covariance matrices using the ALS technique is available online under the GNU General Public License license.

Optimality and Performance

It follows from theory that the Kalman filter is the optimal linear filter in cases where a) the model perfectly matches the real system, b) the entering noise is white (uncorrelated) and c) the covariances of the noise are exactly known. Several methods for the noise covariance estimation have been proposed during past decades, including ALS, mentioned in the section above. After the covariances are estimated, it is useful to evaluate the performance of the filter, i.e. whether it is possible to improve the state estimation quality. If the Kalman filter works optimally, the innovation sequence (the output prediction error) is a white noise, therefore the whiteness property of the innovations measures filter performance. Several different methods can be used for this purpose. If the noise terms are non-Gaussian distributed, methods for assessing performance of the filter estimate, which use probability inequalities or large-sample theory, are given in and.

EXAMPLE APPLICATION, TECHNICAL

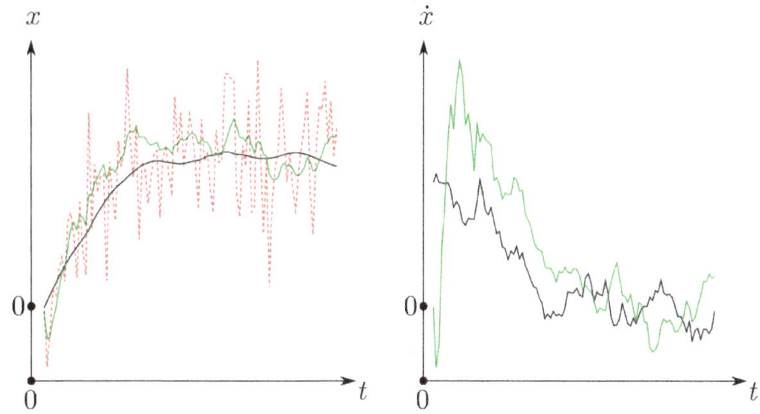

Black: truth, green: filtered process, red: observations

Consider a truck on frictionless, straight rails. Initially, the truck is stationary at position 0, but it is buffeted this way and that by random uncontrolled forces. We measure the position of the truck every Δt seconds, but these measurements are imprecise; we want to maintain a model of where the truck is and what its velocity is. We show here how we derive the model from which we create our Kalman filter.

Since $\mathbf{F}, \mathbf{H}, \mathbf{R}, \mathbf{Q}$ are constant, their time indices are dropped.

The position and velocity of the truck are described by the linear state space

$$\mathbf{x}_k = \begin{bmatrix} x \\ \dot{x} \end{bmatrix}$$

where $\dot{x}$ is the velocity, that is, the derivative of position with respect to time.

We assume that between the (k − 1) and k timestep uncontrolled forces cause a constant acceleration of ak that is normally distributed, with mean 0 and standard deviation σa. From Newton's laws of motion we conclude that

$$\mathbf{x}_k = \mathbf{F}\mathbf{x}_{k-1} + \mathbf{G}a_k$$

(note that there is no Bu term since we have no known control inputs. Instead, we assume that ak is the effect of an unknown input and G applies that effect to the state vector) where

$$\mathbf{F} = \begin{bmatrix} 1 & \Delta t \\ 0 & 1 \end{bmatrix}$$

and

$$\mathbf{G} = \begin{bmatrix} \frac{\Delta t^2}{2} \\ \Delta t \end{bmatrix}$$

so that

$$\mathbf{x}_k = \mathbf{F}\mathbf{x}_{k-1} + \mathbf{w}_k$$

where $\mathbf{w}_k \sim N(0, \mathbf{Q})$ and

$$\mathbf{Q} = \mathbf{G}\mathbf{G}^{\mathrm{T}}\sigma_a^2 = \begin{bmatrix} \frac{\Delta t^4}{4} & \frac{\Delta t^3}{2} \\ \frac{\Delta t^3}{2} & \Delta t^2 \end{bmatrix}\sigma_a^2.$$

0Please note that the matrix Q is not full rank (it is of rank one if $\Delta t \neq 0$). Hence, the distribution 0 $N(0, \mathbf{Q})$ is not absolutely continuous and has no probability density function. Another way to express 0this, avoiding explicit degenerate distributions is given by

0 $\mathbf{w}_k \sim \mathbf{G} \cdot N(0, \sigma_a)$.

A0t each time step, a noisy measurement of the true position of the truck is made. Let us suppose the measurement noise v_k is also normally distributed, with mean 0 and standard deviation σ_z.

0 $\mathbf{z}_k = \mathbf{H}\mathbf{x}_k + \mathbf{v}_k$

w0here

$$\mathbf{H} = [1 \quad 0]$$

and

$$\mathbf{R} = \mathrm{E}[\mathbf{v}_k\mathbf{v}_k^{\mathrm{T}}] = [\sigma_z^2]$$

We know the initial starting state of the truck with perfect precision, so we initialize

$$\hat{\mathbf{x}}_{0|0} = \begin{bmatrix} 0 \\ 0 \end{bmatrix}$$

and to tell the filter that we know the exact position and velocity, we give it a zero covariance matrix:

$$\mathbf{P}_{0|0} = \begin{bmatrix} 0 & 0 \\ 0 & 0 \end{bmatrix}$$

If the initial position and velocity are not known perfectly, the covariance matrix should be initialized with suitable variances on its diagonal:

$$\mathbf{P}_{0|0} = \begin{bmatrix} \sigma_x^2 & 0 \\ 0 & \sigma_{\dot{x}}^2 \end{bmatrix}$$

The filter will then prefer the information from the first measurements over the information already in the model.

DERIVATIONS

Deriving the a posteriori estimate covariance matrix

Starting with our invariant on the error covariance Pk | k as above

$$\mathbf{P}_{k|k} = \mathrm{cov}(\mathbf{x}_k - \hat{\mathbf{x}}_{k|k})$$

substitute in the definition of $\hat{\mathbf{x}}_{k|k}$

$$\mathbf{P}_{k|k} = \mathrm{cov}(\mathbf{x}_k - (\hat{\mathbf{x}}_{k|k-1} + \mathbf{K}_k\tilde{\mathbf{y}}_k))$$

and substitute $\tilde{\mathbf{y}}_k$

$$\mathbf{P}_{k|k} = \mathrm{cov}(\mathbf{x}_k - (\hat{\mathbf{x}}_{k|k-1} + \mathbf{K}_k(\mathbf{z}_k - \mathbf{H}_k\hat{\mathbf{x}}_{k|k-1})))$$

and $\mathbf{z}_k$

$$\mathbf{P}_{k|k} = \mathrm{cov}(\mathbf{x}_k - (\hat{\mathbf{x}}_{k|k-1} + \mathbf{K}_k(\mathbf{H}_k\mathbf{x}_k + \mathbf{v}_k - \mathbf{H}_k\hat{\mathbf{x}}_{k|k-1})))$$

and $\mathbf{z}_k$

$$\mathbf{P}_{k|k} = \mathrm{cov}(\mathbf{x}_k - (\hat{\mathbf{x}}_{k|k-1} + \mathbf{K}_k(\mathbf{H}_k\mathbf{x}_k + \mathbf{v}_k - \mathbf{H}_k\hat{\mathbf{x}}_{k|k-1})))$$

and by collecting the error vectors we get

$$\mathbf{P}_{k|k} = \mathrm{cov}((I - \mathbf{K}_k\mathbf{H}_k)(\mathbf{x}_k - \hat{\mathbf{x}}_{k|k-1}) - \mathbf{K}_k\mathbf{v}_k)$$

Since the measurement error $\mathbf{v}_k$ is uncorrelated with the other terms, this becomes

$$\mathbf{P}_{k|k} = \mathrm{cov}((I - \mathbf{K}_k\mathbf{H}_k)(\mathbf{x}_k - \hat{\mathbf{x}}_{k|k-1})) + \mathrm{cov}(\mathbf{K}_k\mathbf{v}_k)$$

by the properties of vector covariance this becomes

$$\mathbf{P}_{k|k} = (I - \mathbf{K}_k\mathbf{H}_k)\mathrm{cov}(\mathbf{x}_k - \hat{\mathbf{x}}_{k|k-1})(I - \mathbf{K}_k\mathbf{H}_k)^{\mathrm{T}} + \mathbf{K}_k\mathrm{cov}(\mathbf{v}_k)\mathbf{K}_k^{\mathrm{T}}$$

which, using our invariant on $\mathbf{P}_{k\,|\,k-1}$ and the definition of $\mathbf{R}_k$ becomes

$$\mathbf{P}_{k|k} = (I - \mathbf{K}_k\mathbf{H}_k)\mathbf{P}_{k|k-1}(I - \mathbf{K}_k\mathbf{H}_k)^{\mathrm{T}} + \mathbf{K}_k\mathbf{R}_k\mathbf{K}_k^{\mathrm{T}}$$

This formula (sometimes known as the "Joseph form" of the covariance update equation) is valid for any value of Kk. It turns out that if Kk is the optimal Kalman gain, this can be simplified further as shown below.

Kalman Gain Derivation

The Kalman filter is a minimum mean-square error estimator. The error in the a posteriori state estimation is

$$\mathbf{x}_k - \hat{\mathbf{x}}_{k|k}$$

We seek to minimize the expected value of the square of the magnitude of this vector, $\mathrm{E}[\|\mathbf{x}_k - \hat{\mathbf{x}}_{k|k}\|^2]$. This is equivalent to minimizing the trace of the a posteriori estimate covariance matrix $\mathbf{P}_{k|k}$. By expanding out the terms in the equation above and collecting, we get:

$$\mathbf{P}_{k|k} = \mathbf{P}_{k|k-1} - \mathbf{K}_k\mathbf{H}_k\mathbf{P}_{k|k-1} - \mathbf{P}_{k|k-1}\mathbf{H}_k^{\mathrm{T}}\mathbf{K}_k^{\mathrm{T}} + \mathbf{K}_k(\mathbf{H}_k\mathbf{P}_{k|k-1}\mathbf{H}_k^{\mathrm{T}} + \mathbf{R}_k)\mathbf{K}_k^{\mathrm{T}}$$

$$= \mathbf{P}_{k|k-1} - \mathbf{K}_k\mathbf{H}_k\mathbf{P}_{k|k-1} - \mathbf{P}_{k|k-1}\mathbf{H}_k^{\mathrm{T}}\mathbf{K}_k^{\mathrm{T}} + \mathbf{K}_k\mathbf{S}_k\mathbf{K}_k^{\mathrm{T}}$$

The trace is minimized when its matrix derivative with respect to the gain matrix is zero. Using the gradient matrix rules and the symmetry of the matrices involved we find that

$$\frac{\partial \operatorname{tr}(\mathbf{P}_{k|k})}{\partial \mathbf{K}_k} = -2(\mathbf{H}_k\mathbf{P}_{k|k-1})^{\mathrm{T}} + 2\mathbf{K}_k\mathbf{S}_k = 0.$$

Solving this for $\mathbf{K}_k$ yields the Kalman gain:

$$\mathbf{K}_k\mathbf{S}_k = (\mathbf{H}_k\mathbf{P}_{k|k-1})^{\mathrm{T}} = \mathbf{P}_{k|k-1}\mathbf{H}_k^{\mathrm{T}}$$

$$\mathbf{K}_k = \mathbf{P}_{k|k-1}\mathbf{H}_k^{\mathrm{T}}\mathbf{S}_k^{-1}$$

This gain, which is known as the *optimal Kalman gain*, is the one that yields MMSE estimates when used.

Simplification of the a posteriori error covariance formula

The formula used to calculate the a posteriori error covariance can be simplified when the Kalman gain equals the optimal value derived above. Multiplying both sides of our Kalman gain formula on the right by SkKkT, it follows that

$$\mathbf{K}_k\mathbf{S}_k\mathbf{K}_k^{\mathrm{T}} = \mathbf{P}_{k|k-1}\mathbf{H}_k^{\mathrm{T}}\mathbf{K}_k^{\mathrm{T}}$$

Referring back to our expanded formula for the *a posteriori* error covariance,

$$\mathbf{P}_{k|k} = \mathbf{P}_{k|k-1} - \mathbf{K}_k\mathbf{H}_k\mathbf{P}_{k|k-1} - \mathbf{P}_{k|k-1}\mathbf{H}_k^{\mathrm{T}}\mathbf{K}_k^{\mathrm{T}} + \mathbf{K}_k\mathbf{S}_k\mathbf{K}_k^{\mathrm{T}}$$

we find the last two terms cancel out, giving

$$\mathbf{P}_{k|k} = \mathbf{P}_{k|k-1} - \mathbf{K}_k\mathbf{H}_k\mathbf{P}_{k|k-1} = (I - \mathbf{K}_k\mathbf{H}_k)\mathbf{P}_{k|k-1}.$$

This formula is computationally cheaper and thus nearly always used in practice, but is only correct for the optimal gain. If arithmetic precision is unusually low causing problems with numerical stability, or if a non-optimal Kalman gain is deliberately used, this simplification cannot be applied; the a posteriori error covariance formula as derived above (Joseph form) must be used.

SENSITIVITY ANALYSIS

The Kalman filtering equations provide an estimate of the state $\hat{\mathbf{x}}_{k|k}$ and its error covariance $\mathbf{P}_{k|k}$ recursively. The estimate and its quality depend on the system parameters and the noise statistics fed as inputs to the estimator. This section analyzes the effect of uncertainties in the statistical inputs to the filter. In the absence of reliable statistics or the true values of noise covariance matrices $\mathbf{Q}_k$ and $\mathbf{R}_k$, the expression

$$\mathbf{P}_{k|k} = (\mathbf{I} - \mathbf{K}_k\mathbf{H}_k)\mathbf{P}_{k|k-1}(\mathbf{I} - \mathbf{K}_k\mathbf{H}_k)^{\mathrm{T}} + \mathbf{K}_k\mathbf{R}_k\mathbf{K}_k^{\mathrm{T}}$$

no longer provides the actual error covariance. In other words, $\mathbf{P}_{k|k} \neq E[(\mathbf{x}_k - \hat{\mathbf{x}}_{k|k})(\mathbf{x}_k - \hat{\mathbf{x}}_{k|k})^{\mathrm{T}}]$. In most real-time applications, the covariance matrices that are used in designing the Kalman filter are different from the actual (true) noise covariances matrices.This sensitivity analysis describes the behavior of the estimation error covariance when the noise covariances as well as the system matrices $\mathbf{F}_k$ and $\mathbf{H}_k$ that are fed as inputs to the filter are incorrect. Thus, the sensitivity analysis describes the robustness (or sensitivity) of the estimator to misspecified statistical and parametric inputs to the estimator.

This discussion is limited to the error sensitivity analysis for the case of statistical uncertainties. Here the actual noise covariances are denoted by $\mathbf{Q}_k^a$ and $\mathbf{R}_k^a$ respectively, whereas the design values used in the estimator are $\mathbf{Q}_k$ and $\mathbf{R}_k$ respectively. The actual error covariance is denoted by $\mathbf{P}_{k|k}^a$ and $\mathbf{P}_{k|k}$ as computed by the Kalman filter is referred to as the Riccati variable. When $\mathbf{Q}_k \equiv \mathbf{Q}_k^a$ and $\mathbf{R}_k \equiv \mathbf{R}_k^a$, this means that $\mathbf{P}_{k|k} = \mathbf{P}_{k|k}^a$. While computing the actual error covariance using $\mathbf{P}_{k|k}^a = E[(\mathbf{x}_k - \hat{\mathbf{x}}_{k|k})(\mathbf{x}_k - \hat{\mathbf{x}}_{k|k})^{\mathrm{T}}]$, substituting for $\hat{\mathbf{x}}_{k|k}$ and using the fact that $E[\mathbf{w}_k\mathbf{w}_k^{\mathrm{T}}] = \mathbf{Q}_k^a$ and $E[\mathbf{v}_k\mathbf{v}_k^{\mathrm{T}}] = \mathbf{R}_k^a$, results in the following recursive equations for $\mathbf{P}_{k|k}^a$:

$$\mathbf{P}_{k|k-1}^a = \mathbf{F}_k\mathbf{P}_{k-1|k-1}^a\mathbf{F}_k^{\mathrm{T}} + \mathbf{Q}_k^a$$

and

$$\mathbf{P}_{k|k}^a = (\mathbf{I} - \mathbf{K}_k\mathbf{H}_k)\mathbf{P}_{k|k-1}^a(\mathbf{I} - \mathbf{K}_k\mathbf{H}_k)^{\mathrm{T}} + \mathbf{K}_k\mathbf{R}_k^a\mathbf{K}_k^{\mathrm{T}}$$

While computing $\mathbf{P}_{k|k}$, by design the filter implicitly assumes that $E[\mathbf{w}_k\mathbf{w}_k^{\mathrm{T}}] = \mathbf{Q}_k$ and $E[\mathbf{v}_k\mathbf{v}_k^{\mathrm{T}}] = \mathbf{R}_k$. Note that the recursive expressions for $\mathbf{P}_{k|k}^a$ and $\mathbf{P}_{k|k}$ are identical except for the presence of $\mathbf{Q}_k^a$ and $\mathbf{R}_k^a$ in place of the design values $\mathbf{Q}_k$ and $\mathbf{R}_k$ respectively.

Square root form

One problem with the Kalman filter is its numerical stability. If the process noise covariance Qk is small, round-off error often causes a small positive eigenvalue to be computed as a negative number. This renders the numerical representation of the state covariance matrix P indefinite, while its true form is positive-definite.

Positive definite matrices have the property that they have a triangular matrix square root P = S·ST. This can be computed efficiently using the Cholesky factorization algorithm, but more importantly, if the covariance is kept in this form, it can never have a negative diagonal or become asymmetric. An equivalent form, which avoids many of the square root operations required by the matrix square root yet preserves the desirable numerical properties, is the U-D decomposition form, P = U·D·UT, where U is a unit triangular matrix (with unit diagonal), and D is a diagonal matrix.

Between the two, the U-D factorization uses the same amount of storage, and somewhat less computation, and is the most commonly used square root form. (Early literature on the relative efficiency is somewhat misleading, as it assumed that square roots were much more time-consuming than divisions,:69 while on 21-st century computers they are only slightly more expensive.)

Efficient algorithms for the Kalman prediction and update steps in the square root form were developed by G. J. Bierman and C. L. Thornton.

The L·D·LT decomposition of the innovation covariance matrix Sk is the basis for another type of numerically efficient and robust square root filter. The algorithm starts with the LU decomposition as implemented in the Linear Algebra PACKage (LAPACK). These results are further factored into the L·D·LT structure with methods given by Golub and Van Loan (algorithm 4.1.2) for a symmetric nonsingular matrix. Any singular covariance matrix is pivoted so that the first diagonal partition is nonsingular and well-conditioned. The pivoting algorithm must retain any portion of the innovation covariance matrix directly corresponding to observed state-variables Hk·xk|k-1 that are associated with auxiliary observations in yk. The l·d·lt square-root filter requires orthogonalization of the observation vector. This may be done with the inverse square-root of the covariance matrix for the auxiliary variables using Method 2 in Higham (2002, p. 263).

RELATIONSHIP TO RECURSIVE BAYESIAN ESTIMATION

The Kalman filter can be presented as one of the simplest dynamic Bayesian networks. The Kalman filter calculates estimates of the true values of states recursively over time using incoming measurements and a mathematical process model. Similarly, recursive Bayesian estimation calculates estimates of an unknown probability density function (PDF) recursively over time using incoming measurements and a mathematical process model.

In recursive Bayesian estimation, the true state is assumed to be an unobserved Markov process, and the measurements are the observed states of a hidden Markov model (HMM).

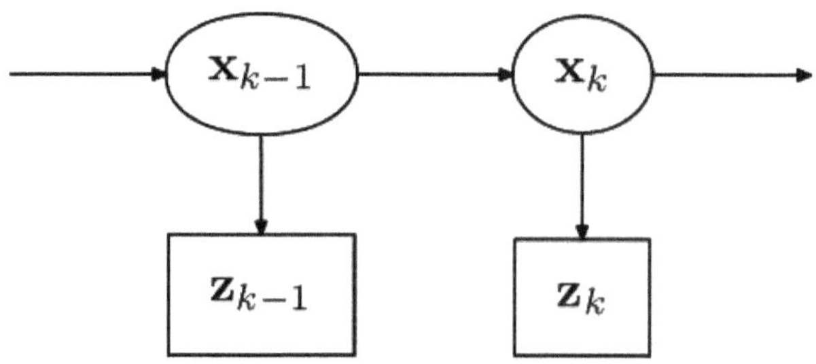

because of the Markov assumption, the true state is conditionally independent of all earlier states given the immediately previous state.

$$p(\mathbf{x}_k \mid \mathbf{x}_0, \ldots, \mathbf{x}_{k-1}) = p(\mathbf{x}_k \mid \mathbf{x}_{k-1})$$

Similarly, the measurement at the *k*-th timestep is dependent only upon the current state and is conditionally independent of all other states given the current state.

$$p(\mathbf{z}_k \mid \mathbf{x}_0, \ldots, \mathbf{x}_k) = p(\mathbf{z}_k \mid \mathbf{x}_k)$$

Using these assumptions the probability distribution over all states of the hidden Markov model can be written simply as:

$$p(\mathbf{x}_0, \ldots, \mathbf{x}_k, \mathbf{z}_1, \ldots, \mathbf{z}_k) = p(\mathbf{x}_0) \prod_{i=1}^{k} p(\mathbf{z}_i \mid \mathbf{x}_i) p(\mathbf{x}_i \mid \mathbf{x}_{i-1})$$

However, when the Kalman filter is used to estimate the state **x**, the probability distribution of interest is that associated with the current states conditioned on the measurements up to the current timestep. This is achieved by marginalizing out the previous states and dividing by the probability of the measurement set.

This leads to the predict and update steps of the Kalman filter written probabilistically. The probability distribution associated with the predicted state is the sum (integral) of the products of the probability distribution associated with the transition from the (k − 1)-th timestep to the k-th and the probability distribution associated with the previous state, over all possible x_{k-1}.

$$p(\mathbf{x}_k \mid \mathbf{Z}_{k-1}) = \int p(\mathbf{x}_k \mid \mathbf{x}_{k-1}) p(\mathbf{x}_{k-1} \mid \mathbf{Z}_{k-1}) \, d\mathbf{x}_{k-1}$$

The measurement set up to time t is

$$\mathbf{Z}_t = \{\mathbf{z}_1, \ldots, \mathbf{z}_t\}$$

The probability distribution of the update is proportional to the product of the measurement likelihood and the predicted state.

$$p(\mathbf{x}_k \mid \mathbf{Z}_k) = \frac{p(\mathbf{z}_k \mid \mathbf{x}_k) p(\mathbf{x}_k \mid \mathbf{Z}_{k-1})}{p(\mathbf{z}_k \mid \mathbf{Z}_{k-1})}$$

The denominator

$$p(\mathbf{z}_k \mid \mathbf{Z}_{k-1}) = \int p(\mathbf{z}_k \mid \mathbf{x}_k) p(\mathbf{x}_k \mid \mathbf{Z}_{k-1}) \, d\mathbf{x}_k$$

is a normalization term.

The remaining probability density functions are

$$p(\mathbf{x}_k \mid \mathbf{x}_{k-1}) = \mathcal{N}(\mathbf{F}_k \mathbf{x}_{k-1}, \mathbf{Q}_k)$$

$$p(\mathbf{z}_k \mid \mathbf{x}_k) = \mathcal{N}(\mathbf{H}_k \mathbf{x}_k, \mathbf{R}_k)$$

$$p(\mathbf{x}_{k-1} \mid \mathbf{Z}_{k-1}) = \mathcal{N}(\hat{\mathbf{x}}_{k-1}, \mathbf{P}_{k-1})$$

Note that the PDF at the previous timestep is inductively assumed to be the estimated state and covariance. This is justified because, as an optimal estimator, the Kalman filter makes best use of the measurements, therefore the PDF for $\mathbf{x}_k$ given the measurements $\mathbf{z}_k$ is the Kalman filter estimate.

Marginal Likelihood

Related to the recursive Bayesian interpretation described above, the Kalman filter can be viewed as a generative model, i.e., a process for generating a stream of random observations z = (z0, z1, z2, ...). Specifically, the process is

- Sample a hidden state X_0 from the Gaussian prior distribution $p(\mathbf{x}_0) = \mathcal{N}(\hat{\mathbf{x}}_{0|0}, \mathbf{P}_{0|0})$
- Sample an observation z_0 from the observation model $p(\mathbf{z}_0 \mid \mathbf{x}_0) = \mathcal{N}(\mathbf{H}_0\mathbf{x}_0, \mathbf{R}_0)$.
- For $k = 1, 2, 3, \ldots$, do
  - Sample the next hidden state $p(\mathbf{x}_k \mid \mathbf{x}_{k-1}) = \mathcal{N}(\mathbf{F}_k\mathbf{x}_{k-1} + \mathbf{B}_k\mathbf{u}_k, \mathbf{Q}_k)$
  - Sample an observation $p(\mathbf{z}_k \mid \mathbf{x}_k) = \mathcal{N}(\mathbf{H}_k\mathbf{x}_k, \mathbf{R}_k)$

Note that this process has identical structure to the hidden Markov model, except that the discrete state and observations are replaced with continuous variables sampled from Gaussian distributions.

In some applications, it is useful to compute the probability that a Kalman filter with a given set of parameters (prior distribution, transition and observation models, and control inputs) would generate a particular observed signal. This probability is known as the marginal likelihood because it integrates over ("marginalizes out") the values of the hidden state variables, so it can be computed using only the observed signal. The marginal likelihood can be useful to evaluate different parameter choices, or to compare the Kalman filter against other models using Bayesian model comparison.

It is straightforward to compute the marginal likelihood as a side effect of the recursive filtering computation. By the chain rule, the likelihood can be factored as the product of the probability of each observation given previous observations,

$$p(\mathbf{z}) = \prod_{k=0}^{T} p(\mathbf{z}_k \mid \mathbf{z}_{k-1}, \ldots, \mathbf{z}_0)$$

and because the Kalman filter describes a Markov process, all relevant information from previous observations is contained in the current state estimate $\hat{\mathbf{x}}_{k|k-1}, \mathbf{P}_{k|k-1}$ Thus the marginal likelihood is given by

$$\begin{aligned}
p(\mathbf{z}) &= \prod_{k=0}^{T} \int p(\mathbf{z}_k \mid \mathbf{x}_k) p(\mathbf{x}_k \mid \mathbf{z}_{k-1}, \ldots, \mathbf{z}_0) d\mathbf{x}_k \\
&= \prod_{k=0}^{T} \int \mathcal{N}(\mathbf{z}_k; \mathbf{H}_k\mathbf{x}_k, \mathbf{R}_k) \mathcal{N}(\mathbf{x}_k; \hat{\mathbf{x}}_{k|k-1}, \mathbf{P}_{k|k-1}) d\mathbf{x}_k \\
&= \prod_{k=0}^{T} \mathcal{N}(\mathbf{z}_k; \mathbf{H}_k\hat{\mathbf{x}}_{k|k-1}, \mathbf{R}_k + \mathbf{H}_k\mathbf{P}_{k|k-1}\mathbf{H}_k^T) \\
&= \prod_{k=0}^{T} \mathcal{N}(\mathbf{z}_k; \mathbf{H}_k\hat{\mathbf{x}}_{k|k-1}, \mathbf{S}_k),
\end{aligned}$$

i.e., a product of Gaussian densities, each corresponding to the density of one observation zk under the current filtering distribution $\mathbf{H}_k\hat{\mathbf{x}}_{k|k-1}, \mathbf{S}_k$. This can easily be computed as a simple recursive update; however, to avoid numeric underflow, in a practical implementation it is usually desirable to compute the log marginal likelihood $\ell = \log p(\mathbf{z})$ instead. Adopting the convention $\ell^{(-1)} = 0$, this can be done via the recursive update rule

$$\ell^{(k)} = \ell^{(k-1)} - \frac{1}{2}\left(\tilde{\mathbf{y}}_k^T \mathbf{S}_k^{-1} \tilde{\mathbf{y}}_k + \log|\mathbf{S}_k| + d_y \log 2\pi\right)$$

where dy is the dimension of the measurement vector.

An important application where such a (log) likelihood of the observations (given the filter parameters) is used is multi-target tracking. For example, consider an object tracking scenario where a stream of observations is the input, however, it is unknown how many objects are in the scene (or, the number of objects is known but is greater than one). In such a scenario, it can be unknown apriori which observations/measurements were generated by which object. A multiple hypothesis tracker (MHT) typically will form different track association hypotheses, where each hypothesis can be viewed as a Kalman filter (in the linear Gaussian case) with a specific set of parameters associated with the hypothesized object. Thus, it is important to compute the likelihood of the observations for the different hypotheses under consideration, such that the most-likely one can be found.

Information Filter

In the information filter, or inverse covariance filter, the estimated covariance and estimated state are replaced by the information matrix and information vector respectively. These are defined as:

$$\mathbf{Y}_{k|k} = \mathbf{P}_{k|k}^{-1}$$
$$\hat{\mathbf{y}}_{k|k} = \mathbf{P}_{k|k}^{-1}\hat{\mathbf{x}}_{k|k}$$

Similarly the predicted covariance and state have equivalent information forms, defined as:

$$\mathbf{Y}_{k|k-1} = \mathbf{P}_{k|k-1}^{-1}$$
$$\hat{\mathbf{y}}_{k|k-1} = \mathbf{P}_{k|k-1}^{-1}\hat{\mathbf{x}}_{k|k-1}$$

as have the measurement covariance and measurement vector, which are defined as:

$$\mathbf{I}_k = \mathbf{H}_k^{\mathrm{T}} \mathbf{R}_k^{-1} \mathbf{H}_k$$
$$\mathbf{i}_k = \mathbf{H}_k^{\mathrm{T}} \mathbf{R}_k^{-1} \mathbf{z}_k$$

The information update now becomes a trivial sum.

$$\mathbf{Y}_{k|k} = \mathbf{Y}_{k|k-1} + \mathbf{I}_k$$
$$\hat{\mathbf{y}}_{k|k} = \hat{\mathbf{y}}_{k|k-1} + \mathbf{i}_k$$

The main advantage of the information filter is that N measurements can be filtered at each timestep simply by summing their information matrices and vectors.

$$\mathbf{Y}_{k|k} = \mathbf{Y}_{k|k-1} + \sum_{j=1}^{N} \mathbf{I}_{k,j}$$

$$\hat{\mathbf{y}}_{k|k} = \hat{\mathbf{y}}_{k|k-1} + \sum_{j=1}^{N} \mathbf{i}_{k,j}$$

To predict the information filter the information matrix and vector can be converted back to their state space equivalents, or alternatively the information space prediction can be used.

$$\mathbf{M}_k = [\mathbf{F}_k^{-1}]^{\mathrm{T}} \mathbf{Y}_{k-1|k-1} \mathbf{F}_k^{-1}$$

$$\mathbf{C}_k = \mathbf{M}_k [\mathbf{M}_k + \mathbf{Q}_k^{-1}]^{-1}$$

$$\mathbf{L}_k = I - \mathbf{C}_k$$

$$\mathbf{Y}_{k|k-1} = \mathbf{L}_k \mathbf{M}_k \mathbf{L}_k^{\mathrm{T}} + \mathbf{C}_k \mathbf{Q}_k^{-1} \mathbf{C}_k^{\mathrm{T}}$$

$$\hat{\mathbf{y}}_{k|k-1} = \mathbf{L}_k [\mathbf{F}_k^{-1}]^{\mathrm{T}} \hat{\mathbf{y}}_{k-1|k-1}$$

Note that if F and Q are time invariant these values can be cached. Note also that F and Q need to be invertible.

FIXED-LAG SMOOTHER

The optimal fixed-lag smoother provides the optimal estimate of $\hat{\mathbf{x}}_{k-N|k}$ for a given fixed-lag N using the measurements from $\mathbf{z}_1$ to $\mathbf{z}_k$. It can be derived using the previous theory via an augmented state, and the main equation of the filter is the following:

$$\begin{bmatrix} \hat{\mathbf{x}}_{t|t} \\ \hat{\mathbf{x}}_{t-1|t} \\ \vdots \\ \hat{\mathbf{x}}_{t-N+1|t} \end{bmatrix} = \begin{bmatrix} \mathbf{I} \\ 0 \\ \vdots \\ 0 \end{bmatrix} \hat{\mathbf{x}}_{t|t-1} + \begin{bmatrix} 0 & \ldots & 0 \\ \mathbf{I} & 0 & \vdots \\ \vdots & \ddots & \vdots \\ 0 & \ldots & I \end{bmatrix} \begin{bmatrix} \hat{\mathbf{x}}_{t-1|t-1} \\ \hat{\mathbf{x}}_{t-2|t-1} \\ \vdots \\ \hat{\mathbf{x}}_{t-N+1|t-1} \end{bmatrix} + \begin{bmatrix} \mathbf{K}^{(0)} \\ \mathbf{K}^{(1)} \\ \vdots \\ \mathbf{K}^{(N-1)} \end{bmatrix} \mathbf{y}_{t|t-1}$$

where:

- $\hat{\mathbf{x}}_{t|t-1}$ is estimated via a standard Kalman filter;
- $\mathbf{y}_{t|t-1} = \mathbf{z}_t - \mathbf{H}\hat{\mathbf{x}}_{t|t-1}$ is the innovation produced considering the estimate of the standard Kalman filter;
- the various $\hat{\mathbf{x}}_{t-i|t}$ with $i = 1, \ldots, N-1$ are new variables, i.e. they do not appear in the standard Kalman filter;
- the gains are computed via the following scheme:

$$\mathbf{K}^{(i)} = \mathbf{P}^{(i)} \mathbf{H}^{T} \left[\mathbf{H}\mathbf{P}\mathbf{H}^{\mathrm{T}} + \mathbf{R}\right]^{-1}$$

and

$$\mathbf{P}^{(i)} = \mathbf{P}\left[[\mathbf{F} - \mathbf{KH}]^T\right]^i$$

where P and K are the prediction error covariance and the gains of the standard Kalman filter $\mathbf{P}_{t|t-1)}$.

If the estimation error covariance is defined so that

$$\mathbf{P}_i := E\left[\left(\mathbf{x}_{t-i} - \hat{\mathbf{x}}_{t-i|t}\right)^* \left(\mathbf{x}_{t-i} - \hat{\mathbf{x}}_{t-i|t}\right) \mid z_1 \ldots z_t\right],$$

then we have that the improvement on the estimation of $\mathbf{x}_{t-i}$ is given by:

$$\mathbf{P} - \mathbf{P}_i = \sum_{j=0}^{i}\left[\mathbf{P}^{(j)}\mathbf{H}^T\left[\mathbf{HPH}^\mathrm{T} + \mathbf{R}\right]^{-1}\mathbf{H}\left(\mathbf{P}^{(i)}\right)^\mathrm{T}\right]$$

FIXED-INTERVAL SMOOTHERS

The optimal fixed-interval smoother provides the optimal estimate of $\hat{\mathbf{x}}_{k|n}$ $(k < n)$ using the measurements from a fixed interval $\mathbf{z}_1$ to $\mathbf{z}_n$. This is also called "Kalman Smoothing". There are several smoothing algorithms in common use.

Rauch–Tung–Striebel

The Rauch–Tung–Striebel (RTS) smoother is an efficient two-pass algorithm for fixed interval smoothing.

The forward pass is the same as the regular Kalman filter algorithm. These filtered a-priori and a-posteriori state estimates $\hat{\mathbf{x}}_{k|k-1}, \hat{\mathbf{x}}_{k|k}$ and covariances $\mathbf{P}_{k|k-1}, \mathbf{P}_{k|k}$ are saved for use in the backwards pass.

In the backwards pass, we compute the smoothed state estimates $\hat{\mathbf{x}}_{k|n}$ and covariances $\mathbf{P}_{k|n}$. We start at the last time step and proceed backwards in time using the following recursive equations:

$$\hat{\mathbf{x}}_{k|n} = \hat{\mathbf{x}}_{k|k} + \mathbf{C}_k(\hat{\mathbf{x}}_{k+1|n} - \hat{\mathbf{x}}_{k+1|k})$$
$$\mathbf{P}_{k|n} = \mathbf{P}_{k|k} + \mathbf{C}_k(\mathbf{P}_{k+1|n} - \mathbf{P}_{k+1|k})\mathbf{C}_k^\mathrm{T}$$

where

$$\mathbf{C}_k = \mathbf{P}_{k|k}\mathbf{F}_{k+1}^\mathrm{T}\mathbf{P}_{k+1|k}^{-1}.$$

Note that $\mathbf{x}_{k|k}$ is the a-posteriori state estimate of timestep k and $\mathbf{x}_{k+1|k}$ is the a-priori state estimate of timestep $k+1$. The same notation applies to the covariance.

Modified Bryson–Frazier smoother

An alternative to the RTS algorithm is the modified Bryson–Frazier (MBF) fixed interval smoother developed by Bierman. This also uses a backward pass that processes data saved from the Kalman filter forward pass. The equations for the backward pass involve the recursive computation of data which are used at each observation time to compute the smoothed state and covariance.

The recursive equations are

$$\tilde{\Lambda}_k = \mathbf{H}_k^T \mathbf{S}_k^{-1} \mathbf{H}_k + \hat{\mathbf{C}}_k^T \hat{\Lambda}_k \hat{\mathbf{C}}_k$$
$$\hat{\Lambda}_{k-1} = \mathbf{F}_k^T \tilde{\Lambda}_k \mathbf{F}_k$$
$$\hat{\Lambda}_n = 0$$
$$\tilde{\lambda}_k = -\mathbf{H}_k^T \mathbf{S}_k^{-1} \mathbf{y}_k + \hat{\mathbf{C}}_k^T \hat{\lambda}_k$$
$$\hat{\lambda}_{k-1} = \mathbf{F}_k^T \tilde{\lambda}_k$$
$$\hat{\lambda}_n = 0$$

where Sk is the residual covariance and $\hat{\mathbf{C}}_k = \mathbf{I} - \mathbf{K}_k \mathbf{H}_k$. The smoothed state and covariance can then be found by substitution in the equations

$$\mathbf{P}_{k|n} = \mathbf{P}_{k|k} - \mathbf{P}_{k|k} \hat{\Lambda}_k \mathbf{P}_{k|k}$$
$$\mathbf{x}_{k|n} = \mathbf{x}_{k|k} - \mathbf{P}_{k|k} \hat{\lambda}_k$$

or

$$\mathbf{P}_{k|n} = \mathbf{P}_{k|k-1} - \mathbf{P}_{k|k-1} \tilde{\Lambda}_k \mathbf{P}_{k|k-1}$$
$$\mathbf{x}_{k|n} = \mathbf{x}_{k|k-1} - \mathbf{P}_{k|k-1} \tilde{\lambda}_k.$$

An important advantage of the MBF is that it does not require finding the inverse of the covariance matrix.

Minimum-variance Smoother

The minimum-variance smoother can attain the best-possible error performance, provided that the models are linear, their parameters and the noise statistics are known precisely. This smoother is a time-varying state-space generalization of the optimal non-causal Wiener filter.

The smoother calculations are done in two passes. The forward calculations involve a one-step-ahead predictor and are given by

$$\hat{\mathbf{x}}_{k+1|k} = (\mathbf{F}_k - \mathbf{K}_k \mathbf{H}_k) \hat{\mathbf{x}}_{k|k-1} + \mathbf{K}_k \mathbf{z}_k$$
$$\alpha_k = -\mathbf{S}_k^{-1/2} \mathbf{H}_k \hat{\mathbf{x}}_{k|k-1} + \mathbf{S}_k^{-1/2} \mathbf{z}_k$$

The above system is known as the inverse Wiener-Hopf factor. The backward recursion is the adjoint of the above forward system. The result of the backward pass β_k may be calculated by operating the forward equations on the time-reversed α_k and time reversing the result. In the case of output estimation, the smoothed estimate is given by

$$\hat{\mathbf{y}}_{k|N} = \mathbf{z}_k - \mathbf{R}_k \beta_k$$

Taking the causal part of this minimum-variance smoother yields

$$\hat{\mathbf{y}}_{k|k} = \mathbf{z}_k - \mathbf{R}_k \mathbf{S}_k^{-1/2} \alpha_k$$

which is identical to the minimum-variance Kalman filter. The above solutions minimize the variance of the output estimation error. Note that the Rauch–Tung–Striebel smoother derivation assumes that the underlying distributions are Gaussian, whereas the minimum-variance solutions do not. Optimal smoothers for state estimation and input estimation can be constructed similarly.

A continuous-time version of the above smoother is described in.

Expectation-maximization algorithms may be employed to calculate approximate maximum likelihood estimates of unknown state-space parameters within minimum-variance filters and smoothers. Often uncertainties remain within problem assumptions. A smoother that accommodates uncertainties can be designed by adding a positive definite term to the Riccati equation.

In cases where the models are nonlinear, step-wise linearizations may be within the minimum-variance filter and smoother recursions (extended Kalman filtering).

FREQUENCY-WEIGHTED KALMAN FILTERS

Pioneering research on the perception of sounds at different frequencies was conducted by Fletcher and Munson in the 1930s. Their work led to a standard way of weighting measured sound levels within investigations of industrial noise and hearing loss. Frequency weightings have since been used within filter and controller designs to manage performance within bands of interest.

Typically, a frequency shaping function is used to weight the average power of the error spectral density in a specified frequency band. Let $\mathbf{y} - \hat{\mathbf{y}}$ denote the output estimation error exhibited by a conventional Kalman filter. Also, let W denote a causal frequency weighting transfer function. The optimum solution which minimizes the variance of $\mathbf{W}(\mathbf{y} - \hat{\mathbf{y}})$ arises by simply constructing $\mathbf{W}^{-1}\hat{\mathbf{y}}$.

The design of W remains an open question. One way of proceeding is to identify a system which generates the estimation error and setting W equal to the inverse of that system. This procedure may be iterated to obtain mean-square error improvement at the cost of increased filter order. The same technique can be applied to smoothers.

Non-Linear Filters

The basic Kalman filter is limited to a linear assumption. More complex systems, however, can be nonlinear. The non-linearity can be associated either with the process model or with the observation model or with both.

Extended Kalman filter

In the extended Kalman filter (EKF), the state transition and observation models need not be linear functions of the state but may instead be non-linear functions. These functions are of differentiable type.

$$\mathbf{x}_k = f(\mathbf{x}_{k-1}, \mathbf{u}_k) + \mathbf{w}_k$$

$$\mathbf{z}_k = h(\mathbf{x}_k) + \mathbf{v}_k$$

The function f can be used to compute the predicted state from the previous estimate and similarly the function h can be used to compute the predicted measurement from the predicted state. However, f and h cannot be applied to the covariance directly. Instead a matrix of partial derivatives (the Jacobian) is computed.

At each timestep the Jacobian is evaluated with current predicted states. These matrices can be used in the Kalman filter equations. This process essentially linearizes the non-linear function around the current estimate.

Unscented Kalman filter

When the state transition and observation models—that is, the predict and update functions f and h— are highly non-linear, the extended Kalman filter can give particularly poor performance. This is because the covariance is propagated through linearization of the underlying non-linear model. The unscented Kalman filter (UKF) uses a deterministic sampling technique known as the unscented transform to pick a minimal set of sample points (called sigma points) around the mean. These sigma points are then propagated through the non-linear functions, from which a new mean and covariance estimate are then formed. The result is a filter which, for certain systems, more accurately estimates the true mean and covariance. This can be verified with Monte Carlo sampling or Taylor series expansion of the posterior statistics. In addition, this technique removes the requirement to explicitly calculate Jacobians, which for complex functions can be a difficult task in itself (i.e., requiring complicated derivatives if done analytically or being computationally costly if done numerically), if not impossible (if those functions are not differentiable).

Predict

As with the EKF, the UKF prediction can be used independently from the UKF update, in combination with a linear (or indeed EKF) update, or vice versa.

The estimated state and covariance are augmented with the mean and covariance of the process noise.

$$\mathbf{x}^a_{k-1|k-1} = [\hat{\mathbf{x}}^T_{k-1|k-1} \quad E[\mathbf{w}^T_k]\,]^T$$

$$\mathbf{P}^a_{k-1|k-1} = \begin{bmatrix} \mathbf{P}_{k-1|k-1} & 0 \\ 0 & \mathbf{Q}_k \end{bmatrix}$$

A set of 2L + 1 sigma points is derived from the augmented state and covariance where L is the dimension of the augmented state.

$$\chi^0_{k-1|k-1} = \mathbf{x}^a_{k-1|k-1}$$
$$\chi^i_{k-1|k-1} = \mathbf{x}^a_{k-1|k-1} + \left(\sqrt{(L+\lambda)\mathbf{P}^a_{k-1|k-1}}\right)_i, \qquad i = 1,\ldots,L$$
$$\chi^i_{k-1|k-1} = \mathbf{x}^a_{k-1|k-1} - \left(\sqrt{(L+\lambda)\mathbf{P}^a_{k-1|k-1}}\right)_{i-L}, \qquad i = L+1,\ldots,2L$$

where

$$\left(\sqrt{(L+\lambda)\mathbf{P}^a_{k-1|k-1}}\right)_i$$

is the *i*th column of the matrix square root of

$$(L+\lambda)\mathbf{P}^a_{k-1|k-1}$$

using the definition: square root A of matrix B satisfies

$$\mathbf{B} \triangleq \mathbf{A}\mathbf{A}^{\mathrm{T}}.$$

The matrix square root should be calculated using numerically efficient and stable methods such as the Cholesky decomposition.

The sigma points are propagated through the transition function *f*.

$$\chi^i_{k|k-1} = f(\chi^i_{k-1|k-1}) \quad i = 0,\ldots,2L$$

where $f : R^L \to R^{|\mathbf{x}|}$. The weighted sigma points are recombined to produce the predicted state and covariance.

$$\hat{\mathbf{x}}_{k|k-1} = \sum_{i=0}^{2L} W^i_s \chi^i_{k|k-1}$$
$$\mathbf{P}_{k|k-1} = \sum_{i=0}^{2L} W^i_c \, [\chi^i_{k|k-1} - \hat{\mathbf{x}}_{k|k-1}][\chi^i_{k|k-1} - \hat{\mathbf{x}}_{k|k-1}]^{\mathrm{T}}$$

where the weights for the state and covariance are given by:

$$W^0_s = \frac{\lambda}{L+\lambda}$$
$$W^0_c = \frac{\lambda}{L+\lambda} + (1 - \alpha^2 + \beta)$$
$$W^i_s = W^i_c = \frac{1}{2(L+\lambda)}$$
$$\lambda = \alpha^2(L+\kappa) - L$$

α and κ control the spread of the sigma points. β is related to the distribution of x. Normal values are $\alpha = 10^{-3}$, $\kappa = 0$ and $\beta = 2$. If the true distribution of x is Gaussian, $\beta = 2$ is optimal.

Update

The predicted state and covariance are augmented as before, except now with the mean and covariance of the measurement noise.

$$\mathbf{x}^a_{k|k-1} = [\hat{\mathbf{x}}^{\mathrm{T}}_{k|k-1} \quad E[\mathbf{v}^{\mathrm{T}}_k]\,]^{\mathrm{T}}$$

$$\mathbf{P}^a_{k|k-1} = \begin{bmatrix} \mathbf{P}_{k|k-1} & 0 \\ 0 & \mathbf{R}_k \end{bmatrix}$$

As before, a set of 2L + 1 sigma points is derived from the augmented state and covariance where L is the dimension of the augmented state.

$$\chi^0_{k|k-1} = \mathbf{x}^a_{k|k-1}$$

$$\chi^i_{k|k-1} = \mathbf{x}^a_{k|k-1} + \left(\sqrt{(L+\lambda)\mathbf{P}^a_{k|k-1}}\right)_i, \qquad i = 1,\ldots,L$$

$$\chi^i_{k|k-1} = \mathbf{x}^a_{k|k-1} - \left(\sqrt{(L+\lambda)\mathbf{P}^a_{k|k-1}}\right)_{i-L}, \qquad i = L+1,\ldots,2L$$

Alternatively if the UKF prediction has been used the sigma points themselves can be augmented along the following lines

$$\chi_{k|k-1} := [\chi^{\mathrm{T}}_{k|k-1} \quad E[\mathbf{v}^{\mathrm{T}}_k]\,]^{\mathrm{T}} \pm \sqrt{(L+\lambda)\mathbf{R}^a_k}$$

where

$$\mathbf{R}^a_k = \begin{bmatrix} 0 & 0 \\ 0 & \mathbf{R}_k \end{bmatrix}$$

The sigma points are projected through the observation function h.

$$\gamma^i_k = h(\chi^i_{k|k-1}) \quad i = 0..2L$$

The weighted sigma points are recombined to produce the predicted measurement and predicted measurement covariance.

$$\hat{\mathbf{z}}_k = \sum_{i=0}^{2L} W^i_s \gamma^i_k$$

$$\mathbf{P}_{z_k z_k} = \sum_{i=0}^{2L} W^i_c\, [\gamma^i_k - \hat{\mathbf{z}}_k][\gamma^i_k - \hat{\mathbf{z}}_k]^{\mathrm{T}}$$

The state-measurement cross-covariance matrix,

$$\mathbf{P}_{x_k z_k} = \sum_{i=0}^{2L} W^i_c\, [\chi^i_{k|k-1} - \hat{\mathbf{x}}_{k|k-1}][\gamma^i_k - \hat{\mathbf{z}}_k]^{\mathrm{T}}$$

is used to compute the UKF Kalman gain.

$$\hat{\mathbf{x}}_{k|k} = \hat{\mathbf{x}}_{k|k-1} + K_k(\mathbf{z}_k - \hat{\mathbf{z}}_k)$$

As with the Kalman filter, the updated state is the predicted state plus the innovation weighted by the Kalman gain,

$$\hat{\mathbf{x}}_{k|k} = \hat{\mathbf{x}}_{k|k-1} + K_k(\mathbf{z}_k - \hat{\mathbf{z}}_k)$$

And the updated covariance is the predicted covariance, minus the predicted measurement covariance, weighted by the Kalman gain.

$$\mathbf{P}_{k|k} = \mathbf{P}_{k|k-1} - K_k \mathbf{P}_{z_k z_k} K_k^{\mathrm{T}}$$

KALMAN–BUCY FILTER

The Kalman–Bucy filter (named after Richard Snowden Bucy) is a continuous time version of the Kalman filter.

It is based on the state space model

$$\frac{d}{dt}\mathbf{x}(t) = \mathbf{F}(t)\mathbf{x}(t) + \mathbf{B}(t)\mathbf{u}(t) + \mathbf{w}(t)$$

$$\mathbf{z}(t) = \mathbf{H}(t)\mathbf{x}(t) + \mathbf{v}(t)$$

where $\mathbf{Q}(t)$ and $\mathbf{R}(t)$ represent the intensities (or, more accurately: the Power Spectral Density - PSD - matrices) of the two white noise terms $\mathbf{w}(t)$ and $\mathbf{v}(t)$, respectively.

The filter consists of two differential equations, one for the state estimate and one for the covariance:

$$\frac{d}{dt}\hat{\mathbf{x}}(t) = \mathbf{F}(t)\hat{\mathbf{x}}(t) + \mathbf{B}(t)\mathbf{u}(t) + \mathbf{K}(t)(\mathbf{z}(t) - \mathbf{H}(t)\hat{\mathbf{x}}(t))$$

$$\frac{d}{dt}\mathbf{P}(t) = \mathbf{F}(t)\mathbf{P}(t) + \mathbf{P}(t)\mathbf{F}^T(t) + \mathbf{Q}(t) - \mathbf{K}(t)\mathbf{R}(t)\mathbf{K}^T(t)$$

where the Kalman gain is given by

$$\mathbf{K}(t) = \mathbf{P}(t)\mathbf{H}^T(t)\mathbf{R}^{-1}(t)$$

Note that in this expression for $\mathbf{K}(t)$ the covariance of the observation noise $\mathbf{R}(t)$ represents at the same time the covariance of the prediction error (or innovation) $\tilde{\mathbf{y}}(t) = \mathbf{z}(t) - \mathbf{H}(t)\hat{\mathbf{x}}(t)$; these covariances are equal only in the case of continuous time.

The distinction between the prediction and update steps of discrete-time Kalman filtering does not exist in continuous time.

The second differential equation, for the covariance, is an example of a Riccati equation.

HYBRID KALMAN FILTER

Most physical systems are represented as continuous-time models while discrete-time measurements are frequently taken for state estimation via a digital processor. Therefore, the system model and measurement model are given by

$$\dot{\mathbf{x}}(t) = \mathbf{F}(t)\mathbf{x}(t) + \mathbf{B}(t)\mathbf{u}(t) + \mathbf{w}(t), \qquad \mathbf{w}(t) \sim N(\mathbf{0}, \mathbf{Q}(t))$$
$$\mathbf{z}_k = \mathbf{H}_k\mathbf{x}_k + \mathbf{v}_k, \qquad \mathbf{v}_k \sim N(\mathbf{0}, \mathbf{R}_k)$$

where

$\mathbf{x}_k = \mathbf{x}(t_k)$.

Initialize

$\hat{\mathbf{x}}_{0|0} = E[\mathbf{x}(t_0)], \mathbf{P}_{0|0} = Var[\mathbf{x}(t_0)]$

Predict

$$\dot{\hat{\mathbf{x}}}(t) = \mathbf{F}(t)\hat{\mathbf{x}}(t) + \mathbf{B}(t)\mathbf{u}(t), \text{ with } \hat{\mathbf{x}}(t_{k-1}) = \hat{\mathbf{x}}_{k-1|k-1}$$
$$\Rightarrow \hat{\mathbf{x}}_{k|k-1} = \hat{\mathbf{x}}(t_k)$$
$$\dot{\mathbf{P}}(t) = \mathbf{F}(t)\mathbf{P}(t) + \mathbf{P}(t)\mathbf{F}(t)^T + \mathbf{Q}(t), \text{ with } \mathbf{P}(t_{k-1}) = \mathbf{P}_{k-1|k-1}$$
$$\Rightarrow \mathbf{P}_{k|k-1} = \mathbf{P}(t_k)$$

The prediction equations are derived from those of continuous-time Kalman filter without update from measurements, i.e., $\mathbf{K}(t) = 0$. The predicted state and covariance are calculated respectively by solving a set of differential equations with the initial value equal to the estimate at the previous step.

Update

$$\mathbf{K}_k = \mathbf{P}_{k|k-1}\mathbf{H}_k^T\left(\mathbf{H}_k\mathbf{P}_{k|k-1}\mathbf{H}_k^T + \mathbf{R}_k\right)^{-1}$$
$$\hat{\mathbf{x}}_{k|k} = \hat{\mathbf{x}}_{k|k-1} + \mathbf{K}_k(\mathbf{z}_k - \mathbf{H}_k\hat{\mathbf{x}}_{k|k-1})$$
$$\mathbf{P}_{k|k} = (\mathbf{I} - \mathbf{K}_k\mathbf{H}_k)\mathbf{P}_{k|k-1}$$

The update equations are identical to those of the discrete-time Kalman filter.

VARIANTS FOR THE RECOVERY OF SPARSE SIGNALS

The traditional Kalman filter has also been employed for the recovery of sparse, possibly dynamic, signals from noisy observations. Recent works utilize notions from the theory of compressed sensing/sampling, such as the restricted isometry property and related probabilistic recovery arguments, for sequentially estimating the sparse state in intrinsically low-dimensional systems.

Applications

- Attitude and heading reference systems
- Autopilot
- Battery state of charge (SoC) estimation
- Brain-computer interface
- Chaotic signals
- Tracking and vertex fitting of charged particles in particle detectors
- Tracking of objects in computer vision

- Dynamic positioning
- Economics, in particular macroeconomics, time series analysis, and econometrics
- Inertial guidance system
- Nuclear medicine – single photon emission computed tomography image restoration
- Orbit Determination
- Power system state estimation
- Radar tracker
- Satellite navigation systems
- Seismology
- Sensorless control of AC motor variable-frequency drives
- Simultaneous localization and mapping
- Speech enhancement
- Visual odometry
- Weather forecasting
- Navigation system
- 3D modeling
- Structural health monitoring
- Human sensorimotor processing

HILBERT TRANSFORM

In mathematics and in signal processing, the Hilbert transform is a linear operator that takes a function, u(t), and produces a function, H(u)(t), with the same domain.

The Hilbert transform is important in signal processing, where it derives the analytic representation of a signal u(t). This means that the real signal u(t) is extended into the complex plane such that it satisfies the Cauchy–Riemann equations. For example, the Hilbert transform leads to the harmonic conjugate of a given function in Fourier analysis, aka harmonic analysis. Equivalently, it is an example of a singular integral operator and of a Fourier multiplier.

The Hilbert transform was originally defined for periodic functions, or equivalently for functions on the circle, in which case it is given by convolution with the Hilbert kernel. More commonly, however, the Hilbert transform refers to a convolution with the Cauchy kernel, for functions defined on the real line R (the boundary of the upper half-plane). The Hilbert transform is closely related to the Paley–Wiener theorem, another result relating holomorphic functions in the upper half-plane and Fourier transforms of functions on the real line.

The Hilbert transform is named after David Hilbert, who first introduced the operator to solve a special case of the Riemann–Hilbert problem for holomorphic functions.

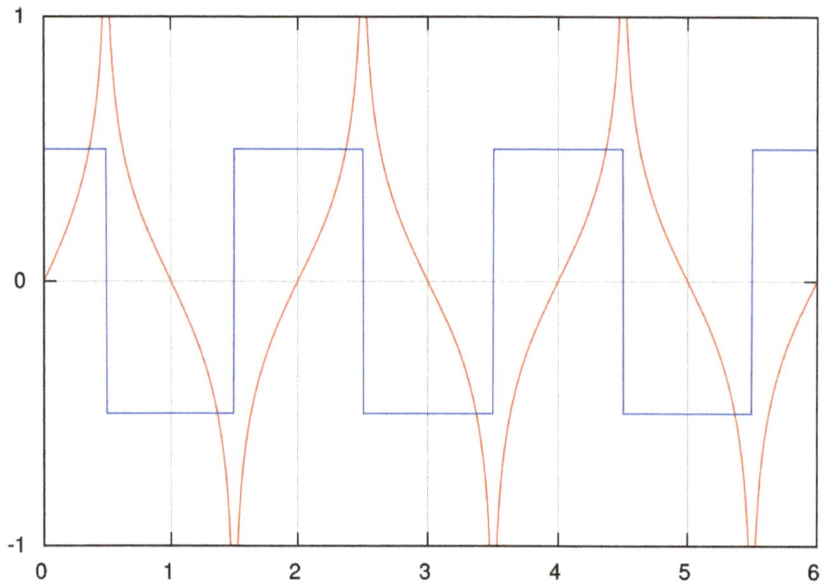

The Hilbert transform, in red, of a square wave, in blue

The Hilbert transform of u can be thought of as the convolution of u(t) with the function h(t) = 1/(πt). Because h(t) is not integrable, the integrals defining the convolution do not converge. Instead, the Hilbert transform is defined using the Cauchy principal value (denoted here by p.v.). Explicitly, the Hilbert transform of a function (or signal) u(t) is given by:

$$H(u)(t) = \text{p. v.} \int_{-\infty}^{\infty} u(\tau)h(t-\tau)\,d\tau = \frac{1}{\pi}\ \text{p. v.} \int_{-\infty}^{\infty} \frac{u(\tau)}{t-\tau}\,d\tau$$

provided this integral exists as a principal value. This is precisely the convolution of u with the tempered distribution p.v. 1/πt (due to Schwartz (1950); see Pandey (1996, Chapter 3)). Alternatively, by changing variables, the principal value integral can be written explicitly (Zygmund 1968, §XVI.1) as:

$$H(u)(t) = -\frac{1}{\pi} \lim_{\varepsilon \to 0} \int_{\varepsilon}^{\infty} \frac{u(t+\tau) - u(t-\tau)}{\tau}\,d\tau.$$

When the Hilbert transform is applied twice in succession to a function *u*, the result is negative *u*:

$$H(H(u))(t) = -u(t)$$

provided the integrals defining both iterations converge in a suitable sense. In particular, the inverse transform is −*H*. This fact can most easily be seen by considering the effect of the Hilbert transform on the Fourier transform of *u*(*t*) (see **Relationship with the Fourier transform**, below).

For an analytic function in upper half-plane, the Hilbert transform describes the relationship between the real part and the imaginary part of the boundary values. That is, if f(z) is analytic in the plane Im z > 0 and u(t) = Re f(t + 0·i) then Im f(t + 0·i) = H(u)(t) up to an additive constant, provided this Hilbert transform exists.

Notation

In signal processing the Hilbert transform of u(t) is commonly denoted by $\widehat{u}(t)$ (e.g., Brandwood 2003, pg 87). However, in mathematics, this notation is already extensively used to denote the Fourier transform of u(t) (e.g., Stein & Weiss 1971). Occasionally, the Hilbert transform may be denoted by $\tilde{u}(t)$. Furthermore, many sources define the Hilbert transform as the negative of the one defined here (e.g., Bracewell 2000, pg 359).

The Hilbert transform arose in Hilbert's 1905 work on a problem Riemann posed concerning analytic functions,(Kress (1989)Bitsadze (2001)) which has come to be known as the Riemann–Hilbert problem. Hilbert's work was mainly concerned with the Hilbert transform for functions defined on the circle (Khvedelidze 2001; Hilbert 1953). Some of his earlier work related to the Discrete Hilbert Transform dates back to lectures he gave in Göttingen. The results were later published by Hermann Weyl in his dissertation (Hardy, Littlewood & Polya 1952, §9.1). Schur improved Hilbert's results about the discrete Hilbert transform and extended them to the integral case (Hardy, Littlewood & Polya 1952, §9.2). These results were restricted to the spaces L2 and ℓ2. In 1928, Marcel Riesz proved that the Hilbert transform can be defined for u in Lp(R) for $1 \leq p < \infty$, that the Hilbert transform is a bounded operator on Lp(R) for $1 < p < \infty$, and that similar results hold for the Hilbert transform on the circle as well as the discrete Hilbert transform (Riesz 1928). The Hilbert transform was a motivating example for Antoni Zygmund and Alberto Calderón during their study of singular integrals (Calderón & Zygmund 1952). Their investigations have played a fundamental role in modern harmonic analysis. Various generalizations of the Hilbert transform, such as the bilinear and trilinear Hilbert transforms are still active areas of research today.

Relationship with the Fourier transform

The Hilbert transform is a multiplier operator (Duoandikoetxea 2000, Chapter 3). The multiplier of H is σH(ω) = −i sgn(ω) where sgn is the signum function. Therefore:

$$\mathcal{F}(H(u))(\omega) = (-i\ \mathrm{sgn}(\omega)) \cdot \mathcal{F}(u)(\omega)$$

where F denotes the Fourier transform. Since sgn(x) = sgn(2πx), it follows that this result applies to the three common definitions of {\displaystyle {\mathcal {F}}}.

By Euler's formula,

$$\sigma_H(\omega) = \begin{cases} i = e^{+\frac{i\pi}{2}}, & \text{for } \omega < 0 \\ 0, & \text{for } \omega = 0 \\ -i = e^{-\frac{i\pi}{2}}, & \text{for } \omega > 0 \end{cases}$$

Therefore, H(u)(t) has the effect of shifting the phase of the negative frequency components of u(t) by +90° (π/2 radians) and the phase of the positive frequency components by −90°. And i·H(u)(t) has the effect of restoring the positive frequency components while shifting the negative frequency ones an additional +90°, resulting in their negation.

When the Hilbert transform is applied twice, the phase of the negative and positive frequency components of $u(t)$ are respectively shifted by +180° and −180°, which are equivalent amounts. The signal is negated; i.e., $H(H(u)) = -u$, because:

$$(\sigma_H(\omega))^2 = e^{\pm i\pi} = -1 \qquad \text{for } \omega \neq 0$$

Table of selected Hilbert transforms

Signal $u(t)$	Hilbert transform[fn 1] $H(u)(t)$
$\sin(t)$ [fn 2]	$-\cos(t)$
$\cos(t)$ [fn 2]	$\sin(t)$
$\exp(it)$	$-i\exp(it)$
$\exp(-it)$	$i\exp(-it)$
$\dfrac{1}{t^2+1}$	$\dfrac{t}{t^2+1}$
e^{-t^2}	$2\pi^{-1/2}F(t)$ see Dawson function
Sinc function $\dfrac{\sin(t)}{t}$	$\dfrac{1-\cos(t)}{t}$
Rectangular function $\sqcap(t)$	$\dfrac{1}{\pi}\ln\left\lvert\dfrac{t+\frac{1}{2}}{t-\frac{1}{2}}\right\rvert$
Dirac delta function $\delta(t)$	$\dfrac{1}{\pi t}$
Characteristic Function $\chi_{[a,b]}(t)$	$\dfrac{1}{\pi}\ln\left\lvert\dfrac{t-a}{t-b}\right\rvert$

An extensive table of Hilbert transforms is available (King 2009). Note that the Hilbert transform of a constant is zero.

Domain of Definition

It is by no means obvious that the Hilbert transform is well-defined at all, as the improper integral defining it must converge in a suitable sense. However, the Hilbert transform is well-defined for a broad class of functions, namely those in Lp(R) for $1 < p < \infty$.

More precisely, if u is in Lp(R) for $1 < p < \infty$, then the limit defining the improper integral

$$H(u)(t) = -\frac{1}{\pi}\lim_{\varepsilon\to 0}\int_{\epsilon}^{\infty}\frac{u(t+\tau)-u(t-\tau)}{\tau}\,d\tau$$

exists for almost every t. The limit function is also in Lp(R) and is in fact the limit in the mean of the improper integral as well. That is,

$$-\frac{1}{\pi}\int_{\epsilon}^{\infty}\frac{u(t+\tau)-u(t-\tau)}{\tau}\,d\tau \to H(u)(t)$$

as ε→0 in the Lp-norm, as well as pointwise almost everywhere, by the Titchmarsh theorem (Titchmarsh 1948, Chapter 5).

In the case p = 1, the Hilbert transform still converges pointwise almost everywhere, but may itself fail to be integrable, even locally (Titchmarsh 1948, §5.14). In particular, convergence in the mean does not in general happen in this case. The Hilbert transform of an L1 function does converge, however, in L1-weak, and the Hilbert transform is a bounded operator from L1 to L1,w (Stein & Weiss 1971, Lemma V.2.8). (In particular, since the Hilbert transform is also a multiplier operator on L2, Marcinkiewicz interpolation and a duality argument furnishes an alternative proof that H is bounded on Lp.)

Properties

Boundedness

If 1 < p < ∞, then the Hilbert transform on Lp(R) is a bounded linear operator, meaning that there exists a constant Cp such that

$$\|Hu\|_p \le C_p\|u\|_p$$

for all u∈Lp(R). This theorem is due to Riesz (1928, VII); see also Titchmarsh (1948, Theorem 101). The best constant Cp is given by

$$C_p = \begin{cases} \tan\frac{\pi}{2p} & \text{for } 1 < p \le 2 \\ \cot\frac{\pi}{2p} & \text{for } 2 < p < \infty \end{cases}$$

This result is due to (Pichorides 1972); see also Grafakos (2004, Remark 4.1.8). The same best constants hold for the periodic Hilbert transform.

The boundedness of the Hilbert transform implies the $L^p(\mathbf{R})$ convergence of the symmetric partial sum operator

$$S_R f = \int_{-R}^{R} \hat{f}(\xi)e^{2\pi i x\xi}\,d\xi$$

to f in Lp(R), see for example (Duoandikoetxea 2000, p. 59).

Anti-self adjointness

The Hilbert transform is an anti-self adjoint operator relative to the duality pairing between Lp(R) and the dual space Lq(R), where p and q are Hölder conjugates and 1 < p,q < ∞. Symbolically,

$$\langle Hu, v\rangle = \langle u, -Hv\rangle$$

for u ∈ Lp(R) and v ∈ Lq(R) (Titchmarsh 1948, Theorem 102).

Inverse transform

The Hilbert transform is an anti-involution (Titchmarsh 1948, p. 120), meaning that

$$H(H(u)) = -u$$

provided each transform is well-defined. Since H preserves the space Lp(R), this implies in particular that the Hilbert transform is invertible on Lp(R), and that

$$H^{-1} = -H$$

Differentiation

Formally, the derivative of the Hilbert transform is the Hilbert transform of the derivative, i.e. these two linear operators commute:

$$H\left(\frac{du}{dt}\right) = \frac{d}{dt}H(u)$$

Iterating this identity,

$$H\left(\frac{d^k u}{dt^k}\right) = \frac{d^k}{dt^k}H(u)$$

This is rigorously true as stated provided u and its first k derivatives belong to Lp(R) (Pandey 1996, §3.3). One can check this easily in the frequency domain, where differentiation becomes multiplication by ω.

Convolutions

The Hilbert transform can formally be realized as a convolution with the tempered distribution (Duistermaat & Kolk 2010, p. 211)

$$h(t) = \text{p. v.} \frac{1}{\pi t}$$

Thus formally,

$$H(u) = h * u$$

However, a priori this may only be defined for u a distribution of compact support. It is possible to work somewhat rigorously with this since compactly supported functions (which are distributions a fortiori) are dense in Lp. Alternatively, one may use the fact that h(t) is the distributional derivative of the function log|t|/π; to wit

$$H(u)(t) = \frac{d}{dt}\left(\frac{1}{\pi}(u * \log|\cdot|)(t)\right)$$

For most operational purposes the Hilbert transform can be treated as a convolution. For example, in a formal sense, the Hilbert transform of a convolution is the convolution of the Hilbert transform on either factor:

$$H(u * v) = H(u) * v = u * H(v)$$

This is rigorously true if u and v are compactly supported distributions since, in that case,

$$h * (u * v) = (h * u) * v = u * (h * v)$$

By passing to an appropriate limit, it is thus also true if u □ Lp and v □ Lr provided

$$1 < \frac{1}{p} + \frac{1}{r}$$

a theorem due to Titchmarsh (1948, Theorem 104).

Invariance

The Hilbert transform has the following invariance properties on $L^2(\mathbf{R})$.

- It commutes with translations. That is, it commutes with the operators $T_a f(x) = f(x + a)$ for all a in **R**
- It commutes with positive dilations. That is it commutes with the operators $M_\lambda f(x) = f(\lambda x)$ for all $\lambda > 0$.
- It anticommutes with the reflection $Rf(x) = f(-x)$.

Up to a multiplicative constant, the Hilbert transform is the only bounded operator on L2 with these properties (Stein 1970, §III.1).

In fact there is a larger group of operators commuting with the Hilbert transform. The group SL(2,R) acts by unitary operators Ug on the space L2(R) by the formula

$$U_g^{-1} f(x) = (cx + d)^{-1} f\left(\frac{ax + b}{cx + d}\right), \quad g = \begin{pmatrix} a & b \\ c & d \end{pmatrix}$$

This unitary representation is an example of a principal series representation of SL(2,R). In this case it is reducible, splitting as the orthogonal sum of two invariant subspaces, Hardy space H2(R) and its conjugate. These are the spaces of L2 boundary values of holomorphic functions on the upper and lower halfplanes. H2(R) and its conjugate consist of exactly those L2 functions with Fourier transforms vanishing on the negative and positive parts of the real axis respectively. Since the Hilbert transform is equal to H = −i (2P − I), with P being the orthogonal projection from L2(R) onto H2(R), it follows that H2(R) and its orthogonal are eigenspaces of H for the eigenvalues ± i. In other words, H commutes with the operators Ug. The restrictions of the operators Ug to H2(R) and its conjugate give irreducible representations of SL(2,R)—the so-called limit of discrete series representations.

Extending the Domain of Definition

Hilbert Transform of Distributions

It is further possible to extend the Hilbert transform to certain spaces of distributions (Pandey 1996, Chapter 3). Since the Hilbert transform commutes with differentiation, and is a bounded operator on Lp, H restricts to give a continuous transform on the inverse limit of Sobolev spaces:

$$\mathcal{D}_{L^p} = \varprojlim_{n\to\infty} W^{n,p}(\mathbb{R})$$

The Hilbert transform can then be defined on the dual space of $\mathcal{D}_{L^p}$, denoted $\mathcal{D}'_{L^p}$, consisting of Lpdistributions. This is accomplished by the duality pairing: for $u \in \mathcal{D}'_{L^p}$, define $H(u) \in \mathcal{D}'_{L^p}$ by

$$\langle Hu, v\rangle \stackrel{\text{def}}{=} \langle u, -Hv\rangle$$

for all $v \in \mathcal{D}_{L^p}$.

It is possible to define the Hilbert transform on the space of tempered distributions as well by an approach due to Gel'fand & Shilov (1967), but considerably more care is needed because of the singularity in the integral.

Hilbert Transform of Bounded Functions

The Hilbert transform can be defined for functions in L∞(R) as well, but it requires some modifications and caveats. Properly understood, the Hilbert transform maps L∞(R) to the Banach space of bounded mean oscillation (BMO) classes.

Interpreted naïvely, the Hilbert transform of a bounded function is clearly ill-defined. For instance, with u = sgn(x), the integral defining H(u) diverges almost everywhere to ±∞. To alleviate such difficulties, the Hilbert transform of an L∞-function is therefore defined by the following regularized form of the integral

$$H(u)(t) = \text{p. v.} \int_{-\infty}^{\infty} u(\tau)\left\{h(t-\tau) - h_0(-\tau)\right\} d\tau$$

where as above $h(x) = 1/\pi x$ and

$$h_0(x) = \begin{cases} 0 & \text{if } |x| < 1 \\ \frac{1}{\pi x} & \text{otherwise} \end{cases}$$

The modified transform H agrees with the original transform on functions of compact support by a general result of Calderón & Zygmund (1952); see Fefferman (1971). The resulting integral, furthermore, converges pointwise almost everywhere, and with respect to the BMO norm, to a function of bounded mean oscillation.

A deep result of Fefferman (1971) and Fefferman & Stein (1972) is that a function is of bounded mean oscillation if and only if it has the form *f* + H(g) for some *f*, g ∈ L∞(R).

Conjugate Functions

The Hilbert transform can be understood in terms of a pair of functions $f(x)$ and $g(x)$ such that the function

$$F(x) = f(x) + ig(x)$$

is the boundary value of a holomorphic function F(z) in the upper half-plane (Titchmarsh 1948, Chapter V). Under these circumstances, if f and g are sufficiently integrable, then one is the Hilbert transform of the other.

Suppose that f ∈ Lp(R). Then, by the theory of the Poisson integral, f admits a unique harmonic extension into the upper half-plane, and this extension is given by

$$u(x+iy) = u(x,y) = \frac{1}{\pi}\int_{-\infty}^{\infty} f(s)\frac{y}{(x-s)^2+y^2}\,ds$$

which is the convolution of f with the Poisson kernel

$$P(x,y) = \frac{1}{\pi}\frac{y}{x^2+y^2}$$

Furthermore, there is a unique harmonic function v defined in the upper half-plane such that F(z) = u(z) + iv(z) is holomorphic and

$$\lim_{y\to\infty} v(x+iy) = 0$$

This harmonic function is obtained from f by taking a convolution with the conjugate Poisson kernel

$$Q(x,y) = \frac{1}{\pi}\frac{x}{x^2+y^2}$$

Thus

$$v(x,y) = \frac{1}{\pi}\int_{-\infty}^{\infty} f(s)\frac{x-s}{(x-s)^2+y^2}\,ds$$

Indeed, the real and imaginary parts of the Cauchy kernel are

$$\frac{i}{\pi z} = P(x,y) + iQ(x,y)$$

so that F = u + iv is holomorphic by Cauchy's integral formula.

The function v obtained from u in this way is called the harmonic conjugate of u. The (non-tangential) boundary limit of v(x,y) as y → 0 is the Hilbert transform of f. Thus, succinctly,

$$H(f) = \lim_{y\to 0} Q(-,y) \star f$$

Titchmarsh's Theorem

A theorem due to Edward Charles Titchmarsh makes precise the relationship between the boundary values of holomorphic functions in the upper half-plane and the Hilbert transform (Titchmarsh 1948, Theorem 95). It gives necessary and sufficient conditions for a complex-valued square-integrable function F(x) on the real line to be the boundary value of a function in the Hardy space H2(U) of holomorphic functions in the upper half-plane U.

The theorem states that the following conditions for a complex-valued square-integrable function F : R → C are equivalent:

- *F*(*x*) is the limit as *z* → *x* of a holomorphic function *F*(*z*) in the upper half-plane such that

$$\int_{-\infty}^{\infty} |F(x+iy)|^2 \, dx < K$$

- The real and imaginary parts of F(x) are Hilbert transforms of each other.
- The Fourier transform $\mathcal{F}(F)(x)$ vanishes for $x < 0$

A weaker result is true for functions of class Lp for p > 1 (Titchmarsh 1948, Theorem 103). Specifically, if F(z) is a holomorphic function such that

$$\int_{-\infty}^{\infty} |F(x+iy)|^p \, dx < K$$

for all *y*, then there is a complex-valued function *F*(*x*) in $L^p(\mathbf{R})$ such that *F*(*x* + i*y*) → *F*(*x*) in the L^p norm as *y* → 0 (as well as holding pointwise almost everywhere). Furthermore,

$$F(x) = f(x) - ig(x)$$

where *f* is a real-valued function in Lp(R) and g is the Hilbert transform (of class Lp) of *f*.

This is not true in the case p = 1. In fact, the Hilbert transform of an L1 function *f* need not converge in the mean to another L1 function. Nevertheless, (Titchmarsh 1948, Theorem 105), the Hilbert transform of *f* does converge almost everywhere to a finite function g such that

$$\int_{-\infty}^{\infty} \frac{|g(x)|^p}{1+x^2} \, dx < \infty$$

This result is directly analogous to one by Andrey Kolmogorov for Hardy functions in the disc (Duren 1970, Theorem 4.2).

Riemann–Hilbert problem

One form of the Riemann–Hilbert problem seeks to identify pairs of functions F+ and F− such that F+ is holomorphic on the upper half-plane and F− is holomorphic on the lower half-plane, such that for x along the real axis,

$$F_+(x) - F_-(x) = f(x)$$

where f(x) is some given real-valued function of x □ R. The left-hand side of this equation may be understood either as the difference of the limits of F± from the appropriate half-planes, or as a hyperfunction distribution. Two functions of this form are a solution of the Riemann–Hilbert problem.

Formally, if F± solve the Riemann–Hilbert problem

$$f(x) = F_+(x) - F_-(x)$$

then the Hilbert transform of $f(x)$ is given by

$$H(f)(x) = \frac{1}{i}(F_+(x) + F_-(x))$$

Hilbert Transform on the Circle

For a periodic function f the circular Hilbert transform is defined as

$$\tilde{f}(x) = \frac{1}{2\pi}\,\text{p.v.}\int_0^{2\pi} f(t)\cot\left(\frac{x-t}{2}\right)dt$$

The circular Hilbert transform is used in giving a characterization of Hardy space and in the study of the conjugate function in Fourier series. The kernel,

$\cot\left(\frac{x-t}{2}\right)$ is known as the **Hilbert kernel** since it was in this form the Hilbert transform was originally studied (Khvedelidze 2001).

The Hilbert kernel (for the circular Hilbert transform) can be obtained by making the Cauchy kernel 1/x periodic. More precisely, for x ≠ 0

$$\frac{1}{2}\cot\left(\frac{x}{2}\right) = \frac{1}{x} + \sum_{n=1}^{\infty}\left(\frac{1}{x+2n\pi} + \frac{1}{x-2n\pi}\right)$$

Many results about the circular Hilbert transform may be derived from the corresponding results for the Hilbert transform from this correspondence.

Another more direct connection is provided by the Cayley transform C(x) = (x – i) / (x + i), which carries the real line onto the circle and the upper half plane onto the unit disk. It induces a unitary map

$$Uf(x) = \pi^{-\frac{1}{2}}(x+i)^{-1}f(C(x))$$

of L2(T) onto L2(R). The operator U carries the Hardy space H2(T) onto the Hardy space H2(R).

Hilbert transform in Signal Processing

Bedrosian's Theorem

Bedrosian's theorem states that the Hilbert transform of the product of a low-pass and a high-pass signal with non-overlapping spectra is given by the product of the low-pass signal and the Hilbert transform of the high-pass signal, or

$$H(f_{LP}(t)f_{HP}(t)) = f_{LP}(t)H(f_{HP}(t))$$

where fLP and fHP are the low- and high-pass signals respectively (Schreier & Scharf 2010, 14).

Amplitude modulated signals are modeled as the product of a bandlimited "message" waveform, um(t), and a sinusoidal "carrier":

$$u(t) = u_m(t) \cdot \cos(\omega t + \phi)$$

When um(t) has no frequency content above the carrier frequency, $\frac{\omega}{2\pi}$ Hz, then by Bedrosian's theorem:

$H(u)(t) = u_m(t) \cdot \sin(\omega t + \phi)$ (Bedrosian 1962)

Analytic Representation

In the context of signal processing, the conjugate function interpretation of the Hilbert transform, discussed above, gives the analytic representation of a signal $u(t)$:

$$u(t) = A \cdot \cos(\omega t + \phi_m(t))$$

which is a holomorphic function in the upper half plane.

For the narrowband model (above), the analytic representation is:

$$\begin{aligned} u_a(t) &= u_m(t) \cdot \cos(\omega t + \phi) + i \cdot u_m(t) \cdot \sin(\omega t + \phi) \\ &= u_m(t) \cdot [\cos(\omega t + \phi) + i \cdot \sin(\omega t + \phi)] \\ &= u_m(t) \cdot e^{i(\omega t + \phi)} \quad \text{(by Euler's formula)} \end{aligned}$$

This complex heterodyne operation shifts all the frequency components of um(t) above 0 Hz. In that case, the imaginary part of the result is a Hilbert transform of the real part. This is an indirect way to produce Hilbert transforms.

Phase/Frequency Modulation

The form:

$$u(t) = A \cdot \cos(\omega t + \phi_m(t))$$

is called phase (or frequency) modulation. The instantaneous frequency is $\omega + \phi_m'(t)$ For sufficiently large ω, compared to ϕ_m':

$$H(u)(t) \approx A \cdot \sin(\omega t + \phi_m(t))$$

and:

$$u_a(t) \approx A \cdot e^{i(\omega t + \phi_m(t))}$$

Single Sideband Modulation (SSB)

When um(t) in Eq.1 is also an analytic representation (of a message waveform), that is:

$$u_m(t) = m(t) + i \cdot \widehat{m}(t)$$

the result is single-sideband modulation:

$$u_a(t) = (m(t) + i \cdot \widehat{m}(t)) \cdot e^{i(\omega t + \phi)}$$

whose transmitted component is:

$$\begin{aligned} u(t) &= \mathrm{Re}\{u_a(t)\} \\ &= m(t) \cdot \cos(\omega t + \phi) - \widehat{m}(t) \cdot \sin(\omega t + \phi) \end{aligned}$$

Causality

The function h with h(t) = 1/(πt) is a non-causal filter and therefore cannot be implemented as is, if u is a time-dependent signal. If u is a function of a non-temporal variable (e.g., spatial) the non-causality might not be a problem. The filter is also of infinite support, which may be a problem in certain applications. Another issue relates to what happens with the zero frequency (DC), which can be avoided by assuring that s does not contain a DC-component.

A practical implementation in many cases implies that a finite support filter, which in addition is made causal by means of a suitable delay, is used to approximate the computation. The approximation may also imply that only a specific frequency range is subject to the characteristic phase shift related to the Hilbert transform. See also quadrature filter.

Discrete Hilbert transform

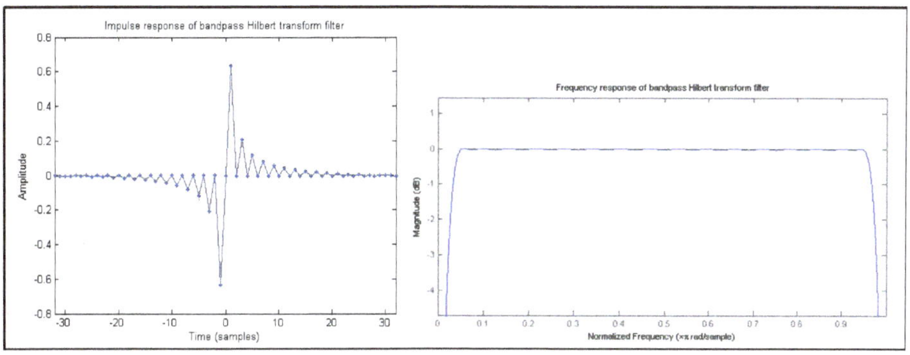

Figure 1: Filter whose frequency response is bandlimited to about 95% of the Nyquist frequency

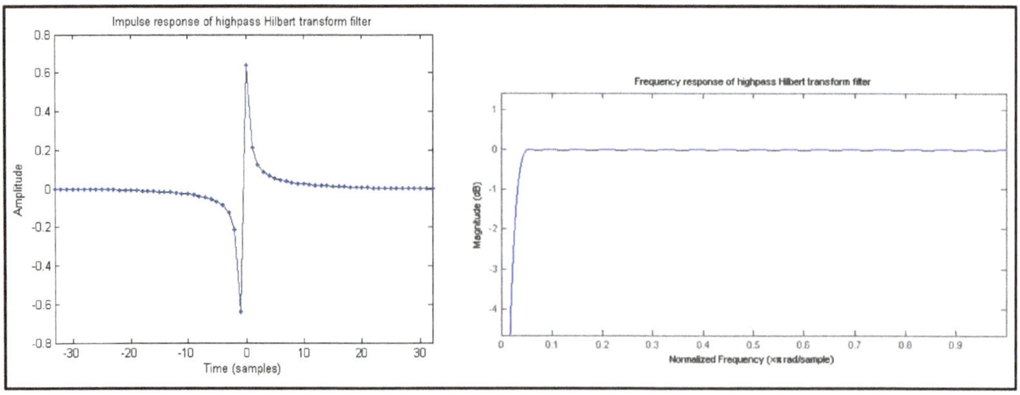

Figure 2: Hilbert transform filter with a highpass frequency response

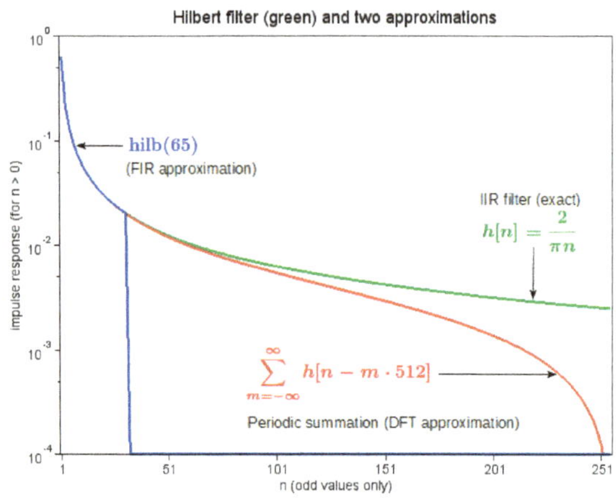

Figure 3.

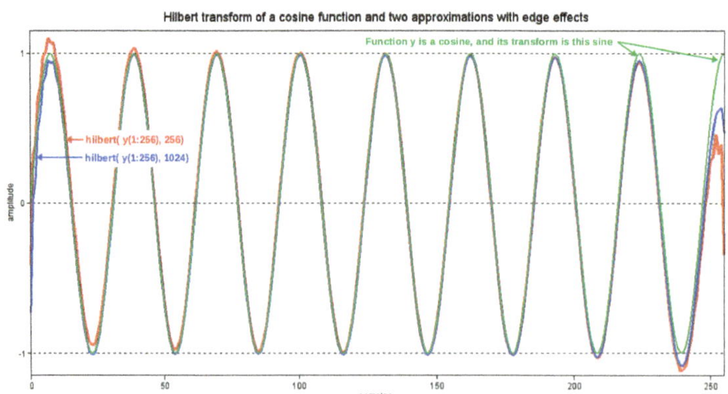

Figure 4: The Hilbert transform of cos(ωt) is sin(ωt). This figure shows sin(ωt) and two approximate Hilbert transforms computed by the MATLAB library function, hilbert(·)

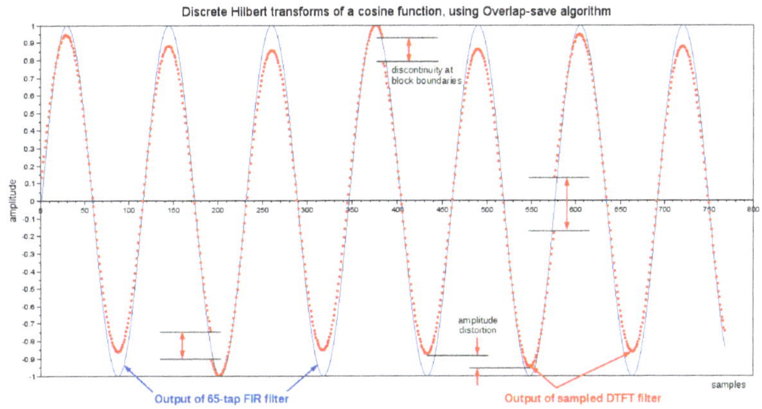

Figure 5. Discrete Hilbert transforms of a cosine function, using piecewise convolution

For a discrete function, u[n] with discrete-time Fourier transform (DTFT), $U(\omega)$ and discrete Hilbert transform $\hat{u}[n]$ the DTFT of $\hat{u}[n]$ in the region $-\pi < \omega < \pi$ is given by: $\mathbf{DTFT}(\hat{u}) = U(\omega) \cdot (-i \cdot \operatorname{sgn}(\omega))$.

The inverse DTFT, using the convolution theorem, is:

$$\begin{aligned}\hat{u}[n] &= \mathit{DTFT}^{-1}(U(\omega)) \ * \ \mathit{DTFT}^{-1}(-i \cdot \operatorname{sgn}(\omega)) \\ &= u[n] \ * \ \frac{1}{2\pi}\int_{-\pi}^{\pi}(-i \cdot \operatorname{sgn}(\omega)) \cdot e^{i\omega n}d\omega \\ &= u[n] \ * \ \underbrace{\frac{1}{2\pi}\left[\int_{-\pi}^{0} i \cdot e^{i\omega n}d\omega - \int_{0}^{\pi} i \cdot e^{i\omega n}d\omega\right]}_{h[n]},\end{aligned}$$

where:

$$h[n] \stackrel{\text{def}}{=} \begin{cases} 0, & \text{for } n \text{ even} \\ \frac{2}{\pi n} & \text{for } n \text{ odd,} \end{cases}$$

which is an infinite impulse response (IIR). When the convolution is performed numerically, an FIR approximation is substituted for h[n], as shown in Figure 1. An FIR filter with an odd number of anti-symmetric coefficients is called Type III, which inherently exhibits responses of zero magnitude at frequencies 0 and Nyquist, resulting in this case in a bandpass filter shape. A Type IV design (even number of anti-symmetric coefficients) is shown in Figure 2. Since the magnitude response at Nyquist does not drop out, it approximates an ideal Hilbert transformer a little better than the odd-tap filter. However:

- A typical (i.e. properly filtered and sampled) $u[n]$ sequence has no useful components at the Nyquist frequency.
- The Type IV impulse response requires a ½ sample shift in the h[n] sequence. That causes the zero-valued coefficients to become non-zero, as seen in Figure 2. So a Type III design is potentially twice as efficient as Type IV.

- The group delay of a Type III design is an integer number of samples, which facilitates aligning [n] with u[n] to create an analytic signal. The group delay of Type IV is halfway between two samples.

The MATLAB function, hilbert(u,N), convolves a u[n] sequence with the periodic summation:

$$h_N[n] \stackrel{\text{def}}{=} \sum_{m=-\infty}^{\infty} h[n-mN],$$

and returns one cycle (N samples) of the periodic result in the imaginary part of a complex-valued output sequence. The convolution is implemented in the frequency domain as the product of the array $DFT(u[n])$ with samples of the −i•sgn(ω) distribution (whose real and imaginary components are all just 0 or ±1). Figure 3 compares a half-cycle of hN[n] with an equivalent length portion of h[n]. Given an FIR approximation for $h[n]$, denoted by $\tilde{h}[n]$, substituting $DFT(\tilde{h}[n])$ for the −i•sgn(ω) samples results in an FIR version of the convolution.

The real part of the output sequence is the original input sequence, so that the complex output is an analytic representation of u[n]. When the input is a segment of a pure cosine, the resulting convolution for two different values of N is depicted in Figure 4 (red and blue plots). Edge effects prevent the result from being a pure sine function (green plot). Since hN[n] is not an FIR sequence, the theoretical extent of the effects is the entire output sequence. But the differences from a sine function diminish with distance from the edges. Parameter N is the output sequence length. If it exceeds the length of the input sequence, the input is modified by appending zero-valued elements. In most cases, that reduces the magnitude of the differences. But their duration is dominated by the inherent rise and fall times of the h[n] impulse response.

An appreciation for the edge effects is important when a method called overlap-save is used to perform the convolution on a long u[n] sequence. Segments of length N are convolved with the periodic function:

$$\tilde{h}_N[n] \stackrel{\text{def}}{=} \sum_{m=-\infty}^{\infty} \tilde{h}[n-mN].$$

When the duration of non-zero values of $\tilde{h}[n]$ is $M < N$, the output sequence includes N − M + 1 samples of $\hat{u}$. $M-1$ outputs are discarded from each block of N, and the input blocks are overlapped by that amount to prevent gaps.

Figure 5 is an example of using both the IIR hilbert() function and the FIR approximation. In the example, a sine function is created by computing the Discrete Hilbert transform of a cosine function, which was processed in four overlapping segments, and pieced back together. As the FIR result (blue) shows, the distortions apparent in the IIR result (red) are not caused by the difference between h[n] and hN[n] (green and red in Fig 3). The fact that hN[n] is tapered (windowed) is actually helpful in this context. The real problem is that it's not windowed enough. Effectively, M = N, whereas the overlap-save method needs M < N.

Number-theoretic Hilbert tRansform

The number theoretic Hilbert transform is an extension (Kak 1970) of the discrete Hilbert transform to integers modulo an appropriate prime number. In this it follows the generalization of discrete Fourier transform to number theoretic transforms. The number theoretic Hilbert transform can be used to generate sets of orthogonal discrete sequences(Kak 2014).

REFERENCES

1. Wolpert, Daniel; Ghahramani, Zoubin (2000). "Computational principles of movement neuroscience". Nature Neuroscience. 3: 1212–7. doi:10.1038/81497. PMID 11127840.
2. Kalman, R. E. (1960). A New Approach to Linear Filtering and Prediction Problems. Journal of Basic Engineering. 82: 35. doi:10.1115/1.3662552.
3. Steffen L. Lauritzen. "Time series analysis in 1880. A discussion of contributions made by T.N. Thiele". International Statistical Review 49, 1981, 319–333. JSTOR 1402616
4. Steffen L. Lauritzen, Thiele: Pioneer in Statistics, Oxford University Press, 2002. ISBN 0-19-850972-3.
5. Mohinder S. Grewal and Angus P. Andrews
6. Gaylor, David; Lightsey, E. Glenn (2003). GPS/INS Kalman Filter Design for Spacecraft Operating in the Proximity of International Space Station. AIAA Guidance, Navigation, and Control Conference and Exhibit. doi:10.2514/6.2003-5445. ISBN 978-1-62410-090-1.
7. Stratonovich, R. L. (1959). Optimum nonlinear systems which bring about a separation of a signal with constant parameters from noise. Radiofizika, 2:6, pp. 892–901.
8. Stratonovich, R. L. (1959). On the theory of optimal non-linear filtering of random functions. Theory of Probability and its Applications, 4, pp. 223–225.
9. Stratonovich, R. L. (1960) Application of the Markov processes theory to optimal filtering. Radio Engineering and Electronic Physics, 5:11, pp. 1–19.
10. Stratonovich, R. L. (1960). Conditional Markov Processes. Theory of Probability and its Applications, 5, pp. 156–178.
11. Ingvar Strid; Karl Walentin (April 2009). Block Kalman Filtering for Large-Scale DSGE Models. Computational Economics. Springer. 33 *(3): 277–304. doi:10.1007/s10614-008-9160-4.*
12. Martin Møller Andreasen (2008). Non-linear DSGE Models, The Central Difference Kalman Filter, and The Mean Shifted Particle Filter (PDF).
13. Roweis, S; Ghahramani, Z (1999). A unifying review of linear gaussian models. Neural computation. 11 *(2): 305–45. doi:10.1162/089976699300016674. PMID 9950734.*
14. Hamilton, J. (1994), Time Series Analysis, Princeton University Press. Chapter 13, 'The Kalman Filter'

15. Ishihara, J.Y.; Terra, M.H.; Campos, J.C.T. (2006). Robust Kalman Filter for Descriptor Systems. IEEE Transactions on Automatic Control. 51 *(8): 1354. doi:10.1109/TAC.2006.878741.*

16. Terra, Marco H.; Cerri, Joao P.; Ishihara, Joao Y. (2014). Optimal Robust Linear Quadratic Regulator for Systems Subject to Uncertainties. IEEE Transactions on Automatic Control. 59 *(9): 2586–2591. doi:10.1109/TAC.2014.2309282.*

17. Kelly, Alonzo (1994). A 3D state space formulation of a navigation Kalman filter for autonomous vehicles (PDF). DTIC Document: 13. 2006 Corrected Version

18. Reid, Ian; Term, Hilary. Estimation II (PDF). www.robots.ox.ac.uk. Oxford University. Retrieved 6 August 2014.

19. Rajamani, Murali (October 2007). Data-based Techniques to Improve State Estimation in Model Predictive Control (PDF) (PhD Thesis). University of Wisconsin–Madison.

20. Rajamani, Murali R.; Rawlings, James B. (2009). Estimation of the disturbance structure from data using semidefinite programming and optimal weighting. Automatica. 45 *(1): 142–148. doi:10.1016/j.automatica.2008.05.032.*

21. Autocovariance Least-Squares Toolbox. Jbrwww.che.wisc.edu. Retrieved 2014-06-02.

22. Three optimality tests with numerical examples are described in Peter, Matisko, (2012). Optimality Tests and Adaptive Kalman Filter. 16th IFAC Symposium on System Identification. 16th IFAC Symposium on System Identification. p. 1523. doi:10.3182/20120711-3-BE-2027.00011. ISBN 978-3-902823-06-9.

23. Spall, James C. (1995). The Kantorovich inequality for error analysis of the Kalman filter with unknown noise distributions. Automatica. 31 *(10): 1513. doi:10.1016/0005-1098(95)00069-9.*

24. Maryak, J.L.; Spall, J.C.; Heydon, B.D. (2004). Use of the Kalman Filter for Inference in State-Space Models with Unknown Noise Distributions. IEEE Transactions on Automatic Control. 49: 87. doi:10.1109/TAC.2003.821415.

25. Anderson, Brian D. O.; Moore, John B. (1979). Optimal Filtering. New York: Prentice Hall. pp. 129–133. ISBN 0-13-638122-7.

26. Thornton, Catherine L. (15 October 1976). Triangular Covariance Factorizations for Kalman Filtering (PDF). (PhD thesis). NASA. NASA Technical Memorandum 33-798.

27. Bierman, G.J. (1977). Factorization Methods for Discrete Sequential Estimation. Factorization Methods for Discrete Sequential Estimation. Academic Press. Bibcode:1977fmds.book.....B.

28. Bar-Shalom, Yaakov; Li, X. Rong; Kirubarajan, Thiagalingam (July 2001). Estimation with Applications to Tracking and Navigation. New York: John Wiley & Sons. pp. 308–317. ISBN 978-0-471-41655-5.

29. Golub, Gene H.; Van Loan, Charles F. (1996). Matrix Computations. Johns Hopkins Studies in the Mathematical Sciences (Third ed.). Baltimore, Maryland: Johns Hopkins University. p. 139. ISBN 978-0-8018-5414-9.

30. Higham, Nicholas J. (2002). Accuracy and Stability of Numerical Algorithms (Second ed.). Philadelphia, PA: Society for Industrial and Applied Mathematics. p. 680. ISBN 978-0-89871-521-7.

31. Masreliez, C. Johan; Martin, R D (1977). Robust Bayesian estimation for the linear

model and robustifying the Kalman filter. IEEE Transactions on Automatic Control. 22 *(3): 361–371. doi:10.1109/TAC.1977.1101538.*

32. Lütkepohl, Helmut (1991). Introduction to Multiple Time Series Analysis. Heidelberg: Springer-Verlag Berlin,. p. 435.
33. Gabriel T. Terejanu (2012-08-04). Discrete Kalman Filter Tutorial (PDF). Retrieved 2016-04-13.
34. Anderson, Brian D. O.; Moore, John B. (1979). Optimal Filtering. Englewood Cliffs, NJ: Prentice Hall, Inc. pp. 176–190. ISBN 0-13-638122-7.
35. Rauch, H.E.; Tung, F.; Striebel, C. T. (August 1965). Maximum likelihood estimates of linear dynamic systems. AIAA Journal. 3 *(8): 1445–1450. Bibcode:1965AIAAJ...3.1445.. doi:10.2514/3.3166.*
36. Einicke, G.A. (March 2006). Optimal and Robust Noncausal Filter Formulations. IEEE Trans. Signal Processing. 54 *(3): 1069–1077. Bibcode:2006ITSP...54.1069E. doi:10.1109/TSP.2005.863042.*
37. Einicke, G.A. (April 2007). Asymptotic Optimality of the Minimum-Variance Fixed-Interval Smoother. IEEE Trans. Signal Processing. 55 *(4): 1543–1547. Bibcode:2007ITSP...55.1543E. doi:10.1109/TSP.2006.889402.*
38. Einicke, G.A.; Ralston, J.C.; Hargrave, C.O.; Reid, D.C.; Hainsworth, D.W. (December 2008). Longwall Mining Automation. An Application of Minimum-Variance Smoothing. IEEE Control Systems Magazine. 28 *(6): 28–37. doi:10.1109/MCS.2008.929281.*
39. Einicke, G.A. (December 2009). Asymptotic Optimality of the Minimum-Variance Fixed-Interval Smoother. IEEE Trans. Automatic Control. 54 *(12): 2904–2908. Bibcode:2007ITSP...55.1543E. doi:10.1109/TSP.2006.889402.*
40. Einicke, G.A. (December 2014). Iterative Frequency-Weighted Filtering and Smoothing Procedures. IEEE Signal Processing Letters. 21 *(12): 1467–1470. Bibcode:2014ISPL...21.1467E. doi:10.1109/LSP.2014.2341641.*
41. Julier, Simon J.; Uhlmann, Jeffrey K. (1997). A new extension of the Kalman filter to nonlinear systems (PDF). Int. Symp. Aerospace/Defense Sensing, Simul. and Controls. Signal Processing, Sensor Fusion, and Target Recognition VI. 3: 182. Bibcode:1997SPIE.3068..182J. doi:10.1117/12.280797. Retrieved 2008-05-03.
42. Gustafsson, Fredrik; Hendeby, Gustaf (2012). Some Relations Between Extended and Unscented Kalman Filters. IEEE Transactions on Signal Processing. 2: 545–555.
43. Wan, E.A.; Van Der Merwe, R. (2000). The unscented Kalman filter for nonlinear estimation. Proceedings of the IEEE 2000 Adaptive Systems for Signal Processing, Communications, and Control Symposium (Cat. No.00EX373) (PDF). p. 153. doi:10.1109/ASSPCC.2000.882463. ISBN 0-7803-5800-7.
44. Bucy, R.S. and Joseph, P.D., Filtering for Stochastic Processes with Applications to Guidance, John Wiley & Sons, 1968; 2nd Edition, AMS Chelsea Publ., 2005. ISBN 0-8218-3782-6
45. Jazwinski, Andrew H., Stochastic processes and filtering theory, Academic Press, New York, 1970. ISBN 0-12-381550-9
46. Kailath, T. (1968). An innovations approach to least-squares estimation--Part I: Linear filtering in additive white noise. IEEE Transactions on Automatic Control. 13 *(6):*

646–655. doi:10.1109/TAC.1968.1099025.

47. Vaswani, Namrata (2008). Kalman filtered Compressed Sensing. 2008 15th IEEE International Conference on Image Processing. p. 893. doi:10.1109/ICIP.2008.4711899. ISBN 978-1-4244-1765-0.

48. Carmi, Avishy; Gurfil, Pini; Kanevsky, Dimitri (2010). Methods for sparse signal recovery using Kalman filtering with embedded pseudo-measurement norms and quasi-norms. IEEE Transactions on Signal Processing. 58 *(4): 2405–2409. Bibcode:2010ITSP...58.2405C. doi:10.1109/TSP.2009.2038959.*

49. Zachariah, Dave; Chatterjee, Saikat; Jansson, Magnus (2012). Dynamic Iterative Pursuit. IEEE Transactions on Signal Processing. 60 *(9): 4967–4972. doi:10.1109/TSP.2012.2203813.*

50. Vasebi, Amir; Partovibakhsh, Maral; Bathaee, S. Mohammad Taghi (2007). A novel combined battery model for state-of-charge estimation in lead-acid batteries based on extended Kalman filter for hybrid electric vehicle applications. Journal of Power Sources. 174: 30–40. doi:10.1016/j.jpowsour.2007.04.011.

51. Vasebi, A.; Bathaee, S.M.T.; Partovibakhsh, M. (2008). Predicting state of charge of lead-acid batteries for hybrid electric vehicles by extended Kalman filter. Energy Conversion and Management. 49: 75–82. doi:10.1016/j.enconman.2007.05.017.

52. Fruhwirth, R. (1987). Application of Kalman filtering to track and vertex fitting. Nucl. Instrum. Meth. A262 *(2–3): 444–450. Bibcode:1987NIMPA.262..444F. doi:10.1016/0168-9002(87)90887-4.*

53. Harvey, Andrew C. (1994). Applications of the Kalman filter in econometrics. In Bewley, Truman. Advances in Econometrics. New York: Cambridge University Press. pp. 285f. ISBN 0-521-46726-8.

54. Boulfelfel, D.; Rangayyan, R.M.; Hahn, L.J.; Kloiber, R.; Kuduvalli, G.R. (1994). Two-dimensional restoration of single photon emission computed tomography images using the Kalman filter. IEEE Transactions on Medical Imaging. 13 *(1): 102–109. doi:10.1109/42.276148. PMID 18218487.*

55. Bock, Y.; Crowell, B.; Webb, F.; Kedar, S.; Clayton, R.; Miyahara, B. (2008). Fusion of High-Rate GPS and Seismic Data: Applications to Early Warning Systems for Mitigation of Geological Hazards. American Geophysical Union. 43: 01. Bibcode:2008AGUFM.G43B..01B.

56. Wolpert, D. M.; Miall, R. C. (1996). Forward Models for Physiological Motor Control. Neural Netw. 9 *(8): 1265–1279. doi:10.1016/S0893-6080(96)00035-4. PMID 12662535.*

CHAPTER 8

Digital Signal Processor

Chapter Contents

- **Introduvtion**
- **Overview**
- **Architecture**
- **History**
- **Modern Dsps**
- **Very Long Instruction Word**
- **Implementations**
- **Transport Triggered Architecture**
- **Mill Architecture**

DSP manipulates different types of signals with the intention of filtering, measuring, or compressing and producing analog signals. Analog signals differ by taking information and translating it into electric pulses of varying amplitude, whereas digital signal information is translated into binary format where each bit of data is represented by two distinguishable amplitudes. Another noticeable difference is that analog signals can be represented as sine waves and digital signals are represented as square waves. DSP can be found in almost any field, whether it's oil processing, sound reproduction, radar and sonar, medical image processing, or telecommunications-- essentially any application in which signals are being compressed and reproduced.

Analog Signal

Digital Signal

INTRODUCTION

A digital signal processor (DSP) is a specialized microprocessor (or a SIP block), with its architecture optimized for the operational needs of digital signal processing.

A digital signal processor chip found in a guitar effects unit.

The goal of DSPs is usually to measure, filter and/or compress continuous real-world analog signals. Most general-purpose microprocessors can also execute digital signal processing algorithms successfully, but dedicated DSPs usually have better power efficiency thus they are more suitable in portable devices such as mobile phones because of power consumption constraints. DSPs often use special memory architectures that are able to fetch multiple data and/or instructions at the same time.

OVERVIEW

A typical digital processing system

Digital signal processing algorithms typically require a large number of mathematical operations to be performed quickly and repeatedly on a series of data samples. Signals (perhaps from audio or video sensors) are constantly converted from analog to digital, manipulated digitally, and then converted back to analog form. Many DSP applications have constraints on latency; that is, for the system to work, the DSP operation must be completed within some fixed time, and deferred (or batch) processing is not viable.

Most general-purpose microprocessors and operating systems can execute DSP algorithms successfully, but are not suitable for use in portable devices such as mobile phones and PDAs because of power efficiency constraints. A specialized digital signal processor, however, will tend to provide a lower-cost solution, with better performance, lower latency, and no requirements for specialized cooling or large batteries.

The architecture of a digital signal processor is optimized specifically for digital signal processing. Most also support some of the features as an applications processor or microcontroller, since signal processing is rarely the only task of a system. Some useful features for optimizing DSP algorithms are outlined below.

ARCHITECTURE

Software Architecture

By the standards of general-purpose processors, DSP instruction sets are often highly irregular; while traditional instruction sets are made up of more general instructions that allow them to perform a wider variety of operations, instruction sets optimized for digital signal processing contain instructions for common mathematical operations that occur frequently in DSP calculations. Both traditional and DSP-optimized instruction sets are able to compute any arbitrary operation but an operation that might require multiple ARM or x86 instructions to compute might require only one instruction in a DSP optimized instruction set.

One implication for software architecture is that hand-optimized assembly-code routines are commonly packaged into libraries for re-use, instead of relying on advanced compiler technologies to handle essential algorithms Even with modern compiler optimizations hand-optimized assembly code is more efficient and many common algorithms involved in DSP calculations are hand-written in order to take full advantage of the architectural optimizations.

Instruction sets

- multiply–accumulates (MACs, including fused multiply–add, FMA) operations
  - used extensively in all kinds of matrix operations
    - convolution for filtering
    - dot product
    - polynomial evaluation

- Fundamental DSP algorithms depend heavily on multiply–accumulate performance
    - FIR filters
    - Fast Fourier transform (FFT)
- Instructions to increase parallelism:
    - SIMD
    - VLIW
    - superscalar architecture
- Specialized instructions for modulo addressing in ring buffers and bit-reversed addressing mode for FFT cross-referencing
- Digital signal processors sometimes use time-stationary encoding to simplify hardware and increase coding efficiency.
- Multiple arithmetic units may require memory architectures to support several accesses per instruction cycle
- Special loop controls, such as architectural support for executing a few instruction words in a very tight loop without overhead for instruction fetches or exit testing

Data Instructions

- Saturation arithmetic, in which operations that produce overflows will accumulate at the maximum (or minimum) values that the register can hold rather than wrapping around (maximum+1 doesn't overflow to minimum as in many general-purpose CPUs, instead it stays at maximum). Sometimes various sticky bits operation modes are available.
- Fixed-point arithmetic is often used to speed up arithmetic processing
- Single-cycle operations to increase the benefits of pipelining

Program Flow

- Floating-point unit integrated directly into the datapath
- Pipelined architecture
- Highly parallel multiplier–accumulators (MAC units)
- Hardware-controlled looping, to reduce or eliminate the overhead required for looping operations

Hardware Architecture

Memory Architecture

DSPs are usually optimized for streaming data and use special memory architectures that are able to fetch multiple data and/or instructions at the same time, such as the Harvard architecture or Modified von Neumann architecture, which use separate program and data memories (sometimes even concurrent access on multiple data buses).

DSPs can sometimes rely on supporting code to know about cache hierarchies and the associated delays. This is a tradeoff that allows for better performance . In addition, extensive use of DMA is employed.

Addressing and Virtual Memory

DSPs frequently use multi-tasking operating systems, but have no support for virtual memory or memory protection. Operating systems that use virtual memory require more time for context switching among processes, which increases latency.

- Hardware modulo addressing
  - Allows circular buffers to be implemented without having to test for wrapping
- Bit-reversed addressing, a special addressing mode
  - useful for calculating FFTs
- Exclusion of a memory management unit
- Memory-address calculation unit

HISTORY

Prior to the advent of stand-alone DSP chips discussed below, most DSP applications were implemented using bit-slice processors. The AMD 2901 bit-slice chip with its family of components was a very popular choice. There were reference designs from AMD, but very often the specifics of a particular design were application specific. These bit slice architectures would sometimes include a peripheral multiplier chip. Examples of these multipliers were a series from TRW including the TDC1008 and TDC1010, some of which included an accumulator, providing the requisite multiply–accumulate (MAC) function.

In 1976, Richard Wiggins proposed the Speak & Spell concept to Paul Breedlove, Larry Brantingham, and Gene Frantz at Texas Instrument's Dallas research facility. Two years later in 1978 they produced the first Speak & Spell, with the technological centerpiece being the TMS5100, the industry's first digital signal processor. It also set other milestones, being the first chip to use Linear predictive coding to perform speech synthesis.

In 1978, Intel released the 2920 as an analog signal processor . It had an on-chip ADC/DAC with an internal signal processor, but it didn't have a hardware multiplier and was not successful in the market. In 1979, AMI released the S2811. It was designed as a microprocessor peripheral, and it had to be initialized by the host. The S2811 was likewise not successful in the market.

In 1980 the first stand-alone, complete DSPs – the NEC μPD7720 and AT&T DSP1 – were presented at the International Solid-State Circuits Conference '80. Both processors were inspired by the research in PSTN telecommunications.

The Altamira DX-1 was another early DSP, utilizing quad integer pipelines with delayed branches and branch prediction.

Another DSP produced by Texas Instruments (TI), the TMS32010 presented in 1983, proved to be an even bigger success. It was based on the Harvard architecture, and so had separate instruction and data memory. It already had a special instruction set, with instructions like load-and-accumulate or multiply-and-accumulate. It could work on 16-bit numbers and needed 390 ns for a multiply–add operation. TI is now the market leader in general-purpose DSPs.

About five years later, the second generation of DSPs began to spread. They had 3 memories for storing two operands simultaneously and included hardware to accelerate tight loops, they also had an addressing unit capable of loop-addressing. Some of them operated on 24-bit variables and a typical model only required about 21 ns for a MAC. Members of this generation were for example the AT&T DSP16A or the Motorola 56000.

The main improvement in the third generation was the appearance of application-specific units and instructions in the data path, or sometimes as coprocessors. These units allowed direct hardware acceleration of very specific but complex mathematical problems, like the Fourier-transform or matrix operations. Some chips, like the Motorola MC68356, even included more than one processor core to work in parallel. Other DSPs from 1995 are the TI TMS320C541 or the TMS 320C80.

The fourth generation is best characterized by the changes in the instruction set and the instruction encoding/decoding. SIMD extensions were added, VLIW and the superscalar architecture appeared. As always, the clock-speeds have increased, a 3 ns MAC now became possible.

MODERN DSPS

Modern signal processors yield greater performance; this is due in part to both technological and architectural advancements like lower design rules, fast-access two-level cache, (E) DMA circuitry and a wider bus system. Not all DSPs provide the same speed and many kinds of signal processors exist, each one of them being better suited for a specific task, ranging in price from about US$1.50 to US$300

Texas Instruments produces the C6000 series DSPs, which have clock speeds of 1.2 GHz and implement separate instruction and data caches. They also have an 8 MiB 2nd level cache and 64 EDMA channels. The top models are capable of as many as 8000 MIPS (instructions per second), use VLIW (very long instruction word), perform eight operations per clock-cycle and are compatible with a broad range of external peripherals and various buses (PCI/serial/etc). TMS320C6474 chips each have three such DSPs, and the newest generation C6000 chips support floating point as well as fixed point processing.

Freescale produces a multi-core DSP family, the MSC81xx. The MSC81xx is based on StarCore Architecture processors and the latest MSC8144 DSP combines four programmable SC3400 StarCore DSP cores. Each SC3400 StarCore DSP core has a clock speed of 1 GHz.

XMOS produces a multi-core multi-threaded line of processor well suited to DSP operations, They come in various speeds ranging from 400 to 1600 MIPS. The processors have a multi-threaded architecture that allows up to 8 real-time threads per core, meaning that a 4 core device would support up to 32 real time threads. Threads communicate between each other with buffered channels that are capable of up to 80 Mbit/s. The devices are easily programmable in C and aim at bridging the gap between conventional micro-controllers and FPGAs

CEVA, Inc. produces and licenses three distinct families of DSPs. Perhaps the best known and most widely deployed is the CEVA-TeakLite DSP family, a classic memory-based architecture, with 16-bit or 32-bit word-widths and single or dual MACs. The CEVA-X DSP family offers a combination of VLIW and SIMD architectures, with different members of the family offering dual or quad 16-bit MACs. The CEVA-XC DSP family targets Software-defined Radio (SDR) modem designs and leverages a unique combination of VLIW and Vector architectures with 32 16-bit MACs.

Analog Devices produce the SHARC-based DSP and range in performance from 66 MHz/198 MFLOPS (million floating-point operations per second) to 400 MHz/2400 MFLOPS. Some models support multiple multipliers and ALUs, SIMD instructions and audio processing-specific components and peripherals. The Blackfin family of embedded digital signal processors combine the features of a DSP with those of a general use processor. As a result, these processors can run simple operating systems like μCLinux, velOSity and Nucleus RTOS while operating on real-time data.

NXP Semiconductors produce DSPs based on TriMedia VLIW technology, optimized for audio and video processing. In some products the DSP core is hidden as a fixed-function block into a SoC, but NXP also provides a range of flexible single core media processors. The TriMedia media processors support both fixed-point arithmetic as well as floating-point arithmetic, and have specific instructions to deal with complex filters and entropy coding.

CSR produces the Quatro family of SoCs that contain one or more custom Imaging DSPs optimized for processing document image data for scanner and copier applications.

Most DSPs use fixed-point arithmetic, because in real world signal processing the additional range provided by floating point is not needed, and there is a large speed benefit and cost benefit due to reduced hardware complexity. Floating point DSPs may be invaluable in applications where a wide dynamic range is required. Product developers might also use floating point DSPs to reduce the cost and complexity of software development in exchange for more expensive hardware, since it is generally easier to implement algorithms in floating point.

Generally, DSPs are dedicated integrated circuits; however DSP functionality can also be produced by using field-programmable gate array chips (FPGAs).

Embedded general-purpose RISC processors are becoming increasingly DSP like in functionality. For example, the OMAP3 processors include a ARM Cortex-A8 and C6000 DSP.

In Communications a new breed of DSPs offering the fusion of both DSP functions and H/W acceleration function is making its way into the mainstream. Such Modem processors include ASOCS ModemX and CEVA's XC4000.

SIMD

Single instruction, multiple data (SIMD), is a class of parallel computers in Flynn's taxonomy. It describes computers with multiple processing elements that perform the same operation on multiple data points simultaneously. Thus, such machines exploit data level parallelism, but not concurrency: there are simultaneous (parallel) computations, but only a single process (instruction) at a given moment. SIMD is particularly applicable to common tasks like adjusting the contrast in a digital image or adjusting the volume of digital audio. Most modern CPU designs include SIMD instructions in order to improve the performance of multimedia use.

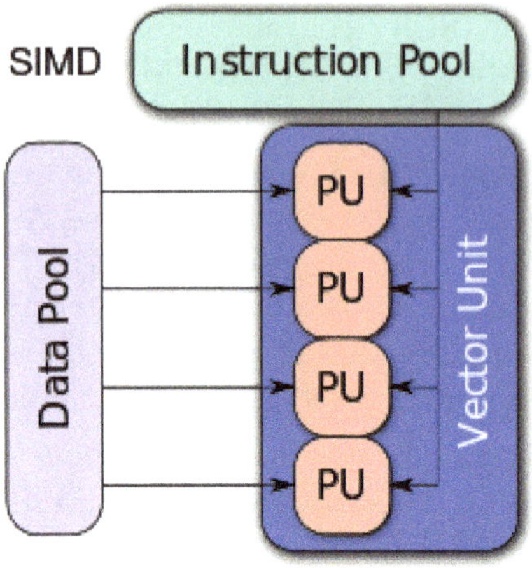

Single instruction, multiple data

The first use of SIMD instructions was in vector supercomputers of the early 1970s such as the CDC Star-100 and the Texas Instruments ASC, which could operate on a vector of data with a single instruction. Vector processing was especially popularized by Cray in the 1970s and 1980s. Vector-processing architectures are now considered separate from SIMD machines, based on the fact that vector machines processed the vectors one word at a time through pipelined processors (though still based on a single instruction), whereas modern SIMD machines process all elements of the vector simultaneously.

The first era of modern SIMD machines was characterized by massively parallel processing-style supercomputers such as the Thinking Machines CM-1 and CM-2. These machines had many limited-functionality processors that would work in parallel. For example, each of 65,536 single-bit processors in a Thinking Machines CM-2 would execute the same instruction at the same time, allowing, for instance, to logically combine 65,536 pairs of bits at a time, using a hypercube-connected network or processor-dedicated RAM to find its operands. Supercomputing moved away from the SIMD approach when inexpensive scalar MIMD approaches based on commodity processors such as the Intel i860 XPbecame more powerful, and interest in SIMD waned.

The current era of SIMD processors grew out of the desktop-computer market rather than the supercomputer market. As desktop processors became powerful enough to support real-time gaming and audio/video processing during the 1990s, demand grew for this particular type of computing power, and microprocessor vendors turned to SIMD to meet the demand. Sun Microsystems introduced SIMD integer instructions in its VIS instruction set extensions in 1995, in its UltraSPARC I microprocessor. MIPS followed suit with their similar MDMX system.

The first widely deployed desktop SIMD was with Intel's MMX extensions to the x86 architecture in 1996. This sparked the introduction of the much more powerful AltiVec system in the Motorola PowerPC's and IBM's POWER systems. Intel responded in 1999 by introducing the all-new SSE system. Since then, there have been several extensions to the SIMD instruction sets for both architectures.

All of these developments have been oriented toward support for real-time graphics, and are therefore oriented toward processing in two, three, or four dimensions, usually with vector lengths of between two and sixteen words, depending on data type and architecture. When new SIMD architectures need to be distinguished from older ones, the newer architectures are then considered short-vector architectures, as earlier SIMD and vector supercomputers had vector lengths from 64 to 64,000. A modern supercomputer is almost always a cluster of MIMD machines, each of which implements (short-vector) SIMD instructions. A modern desktop computer is often a multiprocessor MIMD machine where each processor can execute short-vector SIMD instructions.

Advantages

An application that may take advantage of SIMD is one where the same value is being added to (or subtracted from) a large number of data points, a common operation in many multimedia applications. One example would be changing the brightness of an image. Each pixel of an image consists of three values for the brightness of the red (R), green (G) and blue (B) portions of the color. To change the brightness, the R, G and B values are read from memory, a value is added to (or subtracted from) them, and the resulting values are written back out to memory.

With a SIMD processor there are two improvements to this process. For one the data is understood to be in blocks, and a number of values can be loaded all at once. Instead of a series of instructions saying retrieve this pixel, now retrieve the next pixel , a SIMD processor will have a single instruction that effectively says retrieve n pixels (where n is a number that varies from design to design). For a variety of reasons, this can take much less time than retrieving each pixel individually, as with traditional CPU design.

Another advantage is that the instruction operates on all loaded data in a single operation. In other words, if the SIMD system works by loading up eight data points at once, the add operation being applied to the data will happen to all eight values at the same time. This parallelism is separate from the parallelism provided by a superscalar processor; the eight values are processed in parallel even on a non-superscalar processor, and a superscalar processor may be able to perform multiple SIMD operations in parallel.

Disadvantages

- Not all algorithms can be vectorized easily. For example, a flow-control-heavy task like code parsing may not easily benefit from SIMD; however, it is theoretically possible to vectorize comparisons and *"batch flow"* to target maximal cache optimality, though this technique will require more intermediate state. Note: Batch-pipeline systems (example: GPUs or software rasterization pipelines) are most advantageous for cache control when implemented with SIMD intrinsics, but they are not exclusive to SIMD features. Further complexity may be apparent to avoid dependence within series such as code strings; while independence is required for vectorization.

- It also has large register files which increases power consumption and chip area.
- Currently, implementing an algorithm with SIMD instructions usually requires human labor; most compilers don't generate SIMD instructions from a typical C program, for instance. Vectorization in compilers is an active area of computer science research. (Compare vector processing.)
- Programming with particular SIMD instruction sets can involve numerous low-level challenges.
  - SIMD may have restrictions on data alignment; programmers familiar with one particular architecture may not expect this.
  - Gathering data into SIMD registers and scattering it to the correct destination locations is tricky (sometimes requiring permute operations) and can be inefficient.
  - Specific instructions like rotations or three-operand addition are not available in some SIMD instruction sets.
  - Instruction sets are architecture-specific: some processors lack SIMD instructions entirely, so programmers must provide non-vectorized implementations (or different vectorized implementations) for them.
  - The early MMX instruction set shared a register file with the floating-point stack, which caused inefficiencies when mixing floating-point and MMX code. However, SSE2 corrects this.

Chronology

Examples of SIMD supercomputers (not including vector processors):

- ILLIAC IV, c. 1974
- ICL Distributed Array Processor (DAP), c. 1974
- Burroughs Scientific Processor, c. 1976
- Geometric-Arithmetic Parallel Processor, from Martin Marietta, starting in 1981, continued at Lockheed Martin, then at Teranex and Silicon Optix
- Massively Parallel Processor (MPP), from NASA/Goddard Space Flight Center, c. 1983-1991
- Connection Machine, models 1 and 2 (CM-1 and CM-2), from Thinking Machines Corporation, c. 1985
- MasPar MP-1 and MP-2, c. 1987-1996
- Zephyr DC computer from Wavetracer, c. 1991
- Xplor, from Pyxsys, Inc., c. 2001.

Hardware

Small-scale (64 or 128 bits) SIMD became popular on general-purpose CPUs in the early 1990s and continued through 1997 and later with Motion Video Instructions (MVI) for Alpha. SIMD instructions can be found, to one degree or another, on most CPUs, including the IBM's AltiVec and SPE for PowerPC, HP's PA-RISC Multimedia Acceleration eXtensions (MAX), Intel's MMX and iwMMXt, SSE, SSE2, SSE3SSSE3 and SSE4.x, AMD's 3DNow!, ARC's ARC Video subsystem, SPARC's VIS and VIS2, Sun's MAJC, ARM's NEON technology, MIPS' MDMX (MaDMaX) and MIPS-3D. The IBM, Sony, Toshiba co-developed Cell Processor's SPU's instruction set is heavily SIMD based. NXP founded by Philips developed several SIMD processors named Xetal. The Xetal has 320 16bit processor elements especially designed for vision tasks.

Modern graphics processing units (GPUs) are often wide SIMD implementations, capable of branches, loads, and stores on 128 or 256 bits at a time.

Intel's AVX SIMD instructions now process 256 bits of data at once. Intel's Larrabee prototype microarchitecture includes more than two 512-bit SIMD registers on each of its cores (VPU: Wide Vector Processing Units), and this 512-bit SIMD capability is being continued in Intel's Many Integrated Core Architecture (Intel MIC).

Software

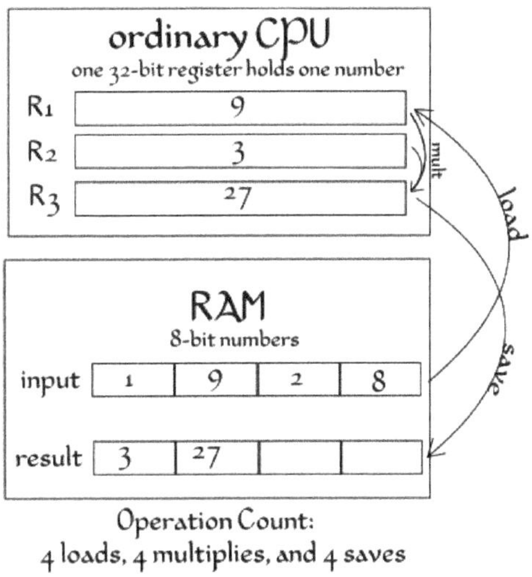

The ordinary tripling of four 8-bit numbers. The CPU loads one 8-bit number into R1, multiplies it with R2, and then saves the answer from R3 back to RAM. This process is repeated for each number.

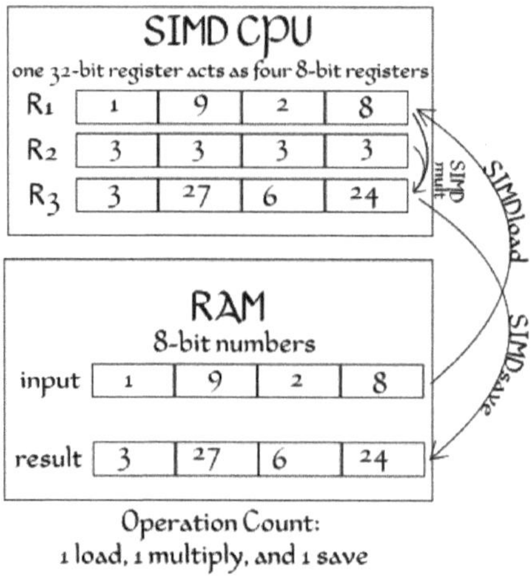

The SIMD tripling of four 8-bit numbers. The CPU loads 4 numbers at once, multiplies them all in one SIMD-multiplication, and saves them all at once back to RAM. In theory, the speed up is about 75%.

SIMD instructions are widely used to process 3D graphics, although modern graphics cards with embedded SIMD have largely taken over this task from the CPU. Some systems also include permute functions that re-pack elements inside vectors, making them particularly useful for data processing and compression. They are also used in cryptography. The trend of general-purpose computing on GPUs (GPGPU) may lead to wider use of SIMD in the future.

Adoption of SIMD systems in personal computer software was at first slow, due to a number of problems. One was that many of the early SIMD instruction sets tended to slow overall performance of the system due to the re-use of existing floating point registers. Other systems, like MMX and 3DNow!, offered support for data types that were not interesting to a wide audience and had expensive context switching instructions to switch between using the FPU and MMX registers. Compilers also often lacked support, requiring programmers to resort to assembly language coding.

SIMD on x86 had a slow start. The introduction of 3DNow! by AMD and SSE by Intel confused matters somewhat, but today the system seems to have settled down (after AMD adopted SSE) and newer compilers should result in more SIMD-enabled software. Intel and AMD now both provide optimized math libraries that use SIMD instructions, and open source alternatives like libSIMD, SIMDx86 and SLEEF have started to appear.

Apple Computer had somewhat more success, even though they entered the SIMD market later than the rest. AltiVec offered a rich system and can be programmed using increasingly sophisticated compilers from Motorola, IBM and GNU, therefore assembly language programming is rarely needed. Additionally, many of the systems that would benefit from SIMD were supplied by Apple itself, for example iTunes and QuickTime. However, in 2006, Apple computers moved to Intel x86 processors. Apple's APIs and development tools (XCode) were modified to support SSE2 and SSE3 as well as AltiVec. Apple was the dominant purchaser of PowerPC chips from IBM and Freescale Semiconductor and even though they abandoned the platform, further development of AltiVec is continued in several Power Architecture designs from Freescale and IBM. On WWDC '15, Apple announced SIMD Vectors support for version 2.0 of their new programming language Swift.

SIMD within a register, or SWAR, is a range of techniques and tricks used for performing SIMD in general-purpose registers on hardware that doesn't provide any direct support for SIMD instructions. This can be used to exploit parallelism in certain algorithms even on hardware that does not support SIMD directly.

Microsoft added SIMD to .NET in RyuJIT. Use of the libraries that implement SIMD on .NET are available in NuGet package Microsoft.Bcl.Simd

SIMD on the Web

In 2013 John McCutchan announced that he had created a performant interface to SIMD instruction sets for the Dart programming language, bringing the benefits of SIMD to web programs for the first time. The interface consists of two types:

- Float32x4, 4 single precision floating point values.
- Int32x4, 4 32-bit integer values.

Instances of these types are immutable and in optimized code are mapped directly to SIMD registers. Operations expressed in Dart typically are compiled into a single instruction with no overhead. This is similar to C and C++ intrinsics. Benchmarks for 4×4 matrix multiplication, 3D vertex transformation, and Mandelbrot set visualization show near 400% speedup compared to scalar code written in Dart.

John's work on Dart has been adopted by ECMAScript and Intel announced at IDF 2013 that they are implementing John's specification for both V8 and SpiderMonkey.

Emscripten, Mozilla's C/C++-to-JavaScript compiler, with extensions can enable compilation of C++ programs that make use of SIMD intrinsics or gcc style vector code to SIMD API of JavaScript resulting in equivalent speedups compared to scalar code.

Commercial Applications

Though it has generally proven difficult to find sustainable commercial applications for SIMD-only processors, one that has had some measure of success is the GAPP, which was developed by Lockheed Martin and taken to the commercial sector by their spin-off Teranex. The GAPP's recent incarnations have become a powerful tool in real-time video processing applications like conversion between various video standards and frame rates (NTSC to/from PAL, NTSC to/from HDTV formats, etc.), deinterlacing, image noise reduction, adaptive video compression, and image enhancement.

A more ubiquitous application for SIMD is found in video games: nearly every modern video game console since 1998 has incorporated a SIMD processor somewhere in its architecture. The PlayStation 2 was unusual in that one of its vector-float units could function as an autonomous DSP executing its own instruction stream, or as a coprocessor driven by ordinary CPU instructions. 3D graphics applications tend to lend themselves well to SIMD processing as they rely heavily on operations with 4-dimensional vectors. Microsoft's Direct3D 9.0 now chooses at runtime processor-specific implementations of its own math operations, including the use of SIMD-capable instructions.

One of the recent processors to use vector processing is the Cell Processor developed by IBM in cooperation with Toshiba and Sony. It uses a number of SIMD processors (a NUMA architecture, each with independent local store and controlled by a general purpose CPU) and is geared towards the huge datasets required by 3D and video processing applications. It differs from traditional ISAs by being SIMD from the ground up with no separate scalar registers.

A recent advancement by Ziilabs was the production of an SIMD type processor which can be used on mobile devices, such as media players and mobile phones.

Larger scale commercial SIMD processors are available from ClearSpeed Technology, Ltd. and Stream Processors, Inc. ClearSpeed's CSX600 (2004) has 96 cores each with 2 double-precision floating point units while the CSX700 (2008) has 192. Stream Processors is headed by computer architect Bill Dally. Their Storm-1 processor (2007) contains 80 SIMD cores controlled by a MIPS CPU.

VERY LONG INSTRUCTION WORD

Very long instruction word (VLIW) refers to processor architectures designed to exploit instruction level parallelism (ILP). Whereas conventional central processing units (CPU, processor) mostly allow programs to specify instructions to execute in sequence only, a VLIW processor allows programs to explicitly specify instructions to execute at the same time, concurrently, in parallel. This design is intended to allow higher performance without the complexity inherent in some other designs.

The traditional means to improve performance in processor architectures include dividing instructions into substeps so the instructions can be executed partly at the same time (termed pipelining), dispatching individual instructions to be executed independently, in different parts of the processor (superscalar architectures), and even executing instructions in an order different from the program (out-of-order execution). These methods all complicate hardware (larger circuits, higher cost and energy use) because the processor must make all of the decisions internally for these methods to work. In contrast, the VLIW method depends on the programs providing all the decisions regarding which instructions to execute simultaneously and how to resolve conflicts. As a practical matter, this means that the compiler (software used to create the final programs) becomes far more complex, but the hardware is simpler than in many other means of parallelism.

The acronym VLIW can also refer to variable-length instruction word, a criterion in instruction set design to allow a more flexible layout of the instruction set and higher code density (depending on the instructions to be used). For example, this method makes it possible to load an immediate value of the size of a machine word into a processor register, which is not feasible if each instruction is limited to the size of a machine word. This flexibility raises instruction decoding needs.

Motivation

A processor that executes every instruction one after the other (i.e., a non-pipelined scalar architecture) may use processor resources inefficiently, yielding potential poor performance. The performance can be improved by executing different substeps of sequential instructions simultaneously (termed pipelining), or even executing multiple instructions entirely simultaneously as in superscalar architectures. Further improvement can be achieved by executing instructions in an order different from that in which they occur in a program, termed out-of-order execution.

These three methods all raise hardware complexity. Before executing any operations in parallel, the processor must verify that the instructions have no interdependencies. For example, if a first instruction's result is used as a second instruction's input, then they cannot execute at the same time and the second instruction cannot execute before the first. Modern out-of-order processors have increased the hardware resources which schedule instructions and determine interdependencies.

In contrast, VLIW executes operations in parallel, based on a fixed schedule, determined when programs are compiled. Since determining the order of execution of operations (including which operations can execute simultaneously) is handled by the compiler, the processor does not need the scheduling hardware that the three methods described above require. Thus, VLIW CPUs offer more computing with less hardware complexity (but greater compiler complexity) than do most superscalar CPUs.

Design

In superscalar designs, the number of execution units is invisible to the instruction set. Each instruction encodes one operation only. For most superscalar designs, the instruction width is 32 bits or fewer.

In contrast, one VLIW instruction encodes multiple operations, at least one operation for each execution unit of a device. For example, if a VLIW device has five execution units, then a VLIW instruction for the device has five operation fields, each field specifying what operation should be done on that corresponding execution unit. To accommodate these operation fields, VLIW instructions are usually at least 64 bits wide, and far wider on some architectures.

For example, the following is an instruction for the Super Harvard Architecture Single-Chip Computer (SHARC). In one cycle, it does a floating-point multiply, a floating-point add, and two autoincrement loads. All of this fits in one 48-bit instruction:

```
f12 = f0 * f4, f8 = f8 + f12, f0 = dm(i0, m3), f4 = pm(i8, m9);
```

Since the earliest days of computer architecture, some CPUs have added several arithmetic logic units (ALUs) to run in parallel. Superscalar CPUs use hardware to decide which operations can run in parallel at runtime, while VLIW CPUs use software (the compiler) to decide which operations can run in parallel in advance. Because the complexity of instruction scheduling is moved into the compiler, complexity of hardware can be reduced substantially

A similar problem occurs when the result of a parallelisable instruction is used as input for a branch. Most modern CPUs guess which branch will be taken even before the calculation is complete, so that they can load the instructions for the branch, or (in some architectures) even start to compute them speculatively. If the CPU guesses wrong, all of these instructions and their context need to be flushed and the correct ones loaded, which takes time.

This has led to increasingly complex instruction-dispatch logic that attempts to guess correctly, and the simplicity of the original reduced instruction set computing (RISC) designs has been eroded. VLIW lacks this logic, and thus lacks its energy use, possible design defects, and other negative aspects.

In a VLIW, the compiler uses heuristics or profile information to guess the direction of a branch. This allows it to move and preschedule operations speculatively before the branch is taken, favoring the most likely path it expects through the branch. If the branch takes an unexpected way, the compiler has already generated compensating code to discard speculative results to preserve program semantics.

Vector processor (single instruction, multiple data (SIMD)) cores can be combined with the VLIW architecture such as in the Fujitsu FR-V microprocessor, further increasing throughput and speed.

The concept of VLIW architecture, and the term VLIW, were invented by Josh Fisher in his research group at Yale University in the early 1980s. His original development of trace scheduling as a compiling method for VLIW was developed when he was a graduate student at New York University. Before VLIW, the notion of prescheduling execution units and instruction-level parallelism in software was well established in the practice of developing horizontal microcode.

Fisher's innovations involved developing a compiler that could target horizontal microcode from programs written in an ordinary programming language. He realized that to get good performance and target a wide-issue machine, it would be necessary to find parallelism beyond that generally within a basic block. He also developed region scheduling methods to identify parallelism beyond basic blocks. Trace scheduling is such a method, and involves scheduling the most likely path of basic blocks first, inserting compensating code to deal with speculative motions, scheduling the second most likely trace, and so on, until the schedule is complete.

Fisher's second innovation was the notion that the target CPU architecture should be designed to be a reasonable target for a compiler; that the compiler and the architecture for a VLIW processor must be codesigned. This was inspired partly by the difficulty Fisher observed at Yale of compiling for architectures like Floating Point Systems' FPS164, which had a complex instruction set computing (CISC) architecture that separated instruction initiation from the instructions that saved the result, needing very complex scheduling algorithms. Fisher developed a set of principles characterizing a proper VLIW design, such as self-draining pipelines, wide multi-port register files, and memory architectures. These principles made it easier for compilers to emit fast code.

The first VLIW compiler was described in a Ph.D. thesis by John Ellis, supervised by Fisher. The compiler was named Bulldog, after Yale's mascot. John Ruttenberg also developed some important algorithms for scheduling.

Fisher left Yale in 1984 to found a startup company, Multiflow, along with cofounders

John O'Donnell and John Ruttenberg. Multiflow produced the TRACE series of VLIW minisupercomputers, shipping their first machines in 1987. Multiflow's VLIW could issue 28 operations in parallel per instruction. The TRACE system was implemented in a mix of medium-scale integration (MSI), large-scale integration (LSI), and very large-scale integration (VLSI), packaged in cabinets, a technology obsoleted as it grew more cost-effective to integrate all of the components of a processor (excluding memory) on one chip.

Multiflow was too early to catch the following wave, when chip architectures began to allow multiple-issue CPUs. The major semiconductor companies recognized the value of Multiflow technology in this context, so the compiler and architecture were subsequently licensed to most of these firms.

IMPLEMENTATIONS

Bob Rau helped develop the Very long instruction word technology

Cydrome was a company producing VLIW numeric processors using emitter-coupled logic (ECL) integrated circuits in the same timeframe (late 1980s). This company, like Multiflow, failed after a few years.

One of the licensees of the Multiflow technology is Hewlett-Packard, which Josh Fisher joined after Multiflow's demise. Bob Rau, founder of Cydrome, also joined HP after Cydrome failed. These two would lead computer architecture research at Hewlett-Packard during the 1990s.

Along with the above systems, during the same time (1989–1990), Intel implemented VLIW in the Intel i860, their first 64-bit microprocessor, and the first processor to implement VLIW on one chip. This processor could operate in both simple RISC mode and VLIW mode:

In the early 1990s, Intel introduced the i860 RISC microprocessor. This simple chip had two modes of operation: a scalar mode and a VLIW mode. In the VLIW mode, the processor always fetched two instructions and assumed that one was an integer instruction and the other floating-point.

The i860's VLIW mode was used extensively in embedded digital signal processor (DSP) applications since the application execution and datasets were simple, well ordered and predictable, allowing designers to fully exploit the parallel execution advantages enabled by VLIW. In VLIW mode, the i860 could maintain floating-point performance in the range of 20-40 double-precision MFLOPS; a very high value for its time and for a processor running at 25-50Mhz.

In the 1990s, Hewlett-Packard researched this problem as a side effect of ongoing work on their PA-RISC processor family. They found that the CPU could be greatly simplified by removing the complex dispatch logic from the CPU and placing it in the compiler. Compilers of the day were far more complex than those of the 1980s, so the added complexity in the compiler was considered to be a small cost.

VLIW CPUs are usually made of multiple RISC-like execution units that operate independently. Contemporary VLIWs usually have four to eight main execution units. Compilers generate initial instruction sequences for the VLIW CPU in roughly the same manner as for traditional CPUs, generating a sequence of RISC-like instructions. The compiler analyzes this code for dependence relationships and resource requirements. It then schedules the instructions according to those constraints. In this process, independent instructions can be scheduled in parallel. Because VLIWs typically represent instructions scheduled in parallel with a longer instruction word that incorporates the individual instructions, this results in a much longer opcode (termed very long) to specify what executes on a given cycle.

Examples of contemporary VLIW CPUs include the TriMedia media processors by NXP (formerly Philips Semiconductors), the Super Harvard Architecture Single-Chip Computer (SHARC) DSP by Analog Devices, the C6000 digital signal processor (DSP) family by Texas Instruments, the ST200 family by STMicroelectronics based on the Lx architecture (designed in Josh Fisher's HP lab by Paolo Faraboschi), and the MPPA Manycore family by Kalray. These contemporary VLIW CPUs are mainly successful as embedded media processors for consumer electronic devices.

VLIW features have also been added to configurable processor cores for system-on-a-chip (SoC) designs. For example, Tensilica's Xtensa LX2 processor incorporates a technology named Flexible Length Instruction eXtensions (FLIX) that allows multi-operation instructions. The Xtensa C/C++ compiler can freely intermix 32- or 64-bit FLIX instructions with the Xtensa processor's one-operation RISC instructions, which are 16 or 24 bits wide. By packing multiple operations into a wide 32- or 64-bit instruction word and allowing these multi-operation instructions to intermix with shorter RISC instructions, FLIX allows SoC designers to realize VLIW's performance advantages while eliminating the code bloat of early VLIW architectures. The Infineon Carmel DSP is another VLIW processor core intended for SoC. It uses a similar code density improvement method called configurable long instruction word (CLIW).

Outside embedded processing markets, Intel's Itanium IA-64 explicitly parallel instruction computing (EPIC) and Elbrus 2000 appear as the only examples of a widely used VLIW CPU architectures. However, EPIC architecture is sometimes distinguished from a pure VLIW architecture, since EPIC advocates full instruction predication, rotating register files, and a very long instruction word that can encode non-parallel instruction groups. VLIWs also gained significant consumer penetration in the graphics processing unit (GPU) market, though both Nvidia and AMD have since moved to RISC architectures to improve performance on non-graphics workloads.

ATI Technologies' (ATI) and Advanced Micro Devices' (AMD) TeraScale microarchitecture for graphics processing units (GPUs) is a VLIW microarchitecture.

In December 2015, the first shipment of PCs based on VLIW CPU Elbrus-4s was made in Russia.

Backward Compatibility

When silicon technology allowed for wider implementations (with more execution units) to be built, the compiled programs for the earlier generation would not run on the wider implementations, as the encoding of binary instructions depended on the number of execution units of the machine.

Transmeta addressed this issue by including a binary-to-binary software compiler layer (termed code morphing) in their Crusoe implementation of the x86 architecture. This mechanism was advertised to basically recompile, optimize, and translate x86 opcodes at runtime into the CPU's internal machine code. Thus, the Transmeta chip is internally a VLIW processor, effectively decoupled from the x86 CISC instruction set that it executes.

Intel's Itanium architecture (among others) solved the backward-compatibility problem with a more general mechanism. Within each of the multiple-opcode instructions, a bit field is allocated to denote dependency on the prior VLIW instruction within the program instruction stream. These bits are set at compile time, thus relieving the hardware from calculating this dependency information. Having this dependency information encoded in the instruction stream allows wider implementations to issue multiple non-dependent VLIW instructions in parallel per cycle, while narrower implementations would issue a smaller number of VLIW instructions per cycle.

Another perceived deficiency of VLIW designs is the code bloat that occurs when one or more execution unit(s) have no useful work to do and thus must execute No Operation NOP instructions. This occurs when there are dependencies in the code and the instruction pipelines must be allowed to drain before later operations can proceed.

Since the number of transistors on a chip has grown, the perceived disadvantages of the VLIW have diminished in importance. VLIW architectures are growing in popularity, especially in the embedded system market, where it is possible to customize a processor for an application in a system-on-a-chip. Embedded VLIW products are available from several vendors, including the FR-V from Fujitsu, the BSP15/16 from Pixelworks, the ST231 from STMicroelectronics, the TriMedia from NXP Semiconductors, the CEVA-X DSP from CEVA, the Jazz DSP from Improv Systems, and Silicon Hive.The Texas Instruments TMS320 DSP line has evolved, in its C6xxx family, to look more like a VLIW, in contrast to the earlier C5xxx family.

SPMD

In computing, SPMD (single program, multiple data) is a technique employed to achieve parallelism; it is a subcategory of MIMD. Tasks are split up and run simultaneously on multiple processors with different input in order to obtain results faster. SPMD is the most common style of parallel programming. It is also a prerequisite for research concepts such as active messages and distributed shared memory.

SPMD vs SIMD

In SPMD, multiple autonomous processors simultaneously execute the same program at independent points, rather than in the lockstep that SIMD imposes on different data. With SPMD, tasks can be executed on general purpose CPUs; SIMD requires vector processors to manipulate data streams. Note that the two are not mutually exclusive.

Distributed Memory

SPMD usually refers to message passing programming on distributed memory computer architectures. A distributed memory computer consists of a collection of independent computers, called nodes. Each node starts its own program and communicates with other nodes by sending and receiving messages, calling send/receive routines for that purpose. Barrier synchronization may also be implemented by messages. The messages can be sent by a number of communication mechanisms, such as TCP/IP over Ethernet, or specialized high-speed interconnects such as Myrinet and Supercomputer Interconnect. Serial sections of the program are implemented by identical computation on all nodes rather than computing the result on one node and sending it to the others.

Nowadays, the programmer is isolated from the details of the message passing by standard interfaces, such as PVM and MPI.

Distributed memory is the programming style used on parallel supercomputers from homegrown Beowulf clusters to the largest clusters on the Teragrid.

Shared Memory

On a shared memory machine (a computer with several CPUs that access the same memory space), messages can be sent by depositing their contents in a shared memory area. This is often the most efficient way to program shared memory computers with large number of processors, especially on NUMA machines, where memory is local to processors and accessing memory of another processor takes longer. SPMD on a shared memory machine is usually implemented by standard (heavyweight) processes.

Unlike SPMD, shared memory multiprocessing, also called symmetric multiprocessing or SMP, presents the programmer with a common memory space and the possibility to parallelize execution by having the program take different paths on different processors. The program starts executing on one processor and the execution splits in a parallel region, which is started when parallel directives are encountered. In a parallel region, the processors execute a single program on different data. A typical example is the parallel DO loop, where different processors work on separate parts of the arrays involved in the loop. At the end of the loop, execution is synchronized, only one processor continues, and the others wait. The current standard interface for shared memory multiprocessing is OpenMP. It is usually implemented by lightweight processes, called threads.

Combination of levels of parallelism

Current computers allow exploiting of many parallel modes at the same time for maximum combined effect. A distributed memory program using MPI may run on a collection of nodes. Each node may be a shared memory computer and execute in parallel on multiple CPUs using OpenMP. Within each CPU, SIMD vector instructions (usually generated automatically by the compiler) and superscalar instruction execution (usually handled transparently by the CPU itself), such as pipelining and the use of multiple parallel functional units, are used for maximum single CPU speed.

History

SPMD was proposed first in 1983 by Michel Auguin (University of Nice Sophia-Antipolis) and François Larbey (Thomson/Sintra) in the OPSILA parallel computer and next in 1984 by Frederica Darema at IBM for highly parallel machines like the RP3 (the IBM Research Parallel Processor Prototype), in an unpublished IBM memo. By the late 1980s, there were many distr

TRANSPORT TRIGGERED ARCHITECTURE

In computer architecture, a transport triggered architecture (TTA) is a kind of CPU design in which programs directly control the internal transport buses of a processor. Computation happens as a side effect of data transports: writing data into a triggering port of a functional unit triggers the functional unit to start a computation. This is similar to what happens in a systolic array. Due to its modular structure, TTA is an ideal processor template for application-specific instruction-set processors (ASIP) with customized datapath but without the inflexibility and design cost of fixed function hardware accelerators.

Typically a transport triggered processor has multiple transport buses and multiple functional units connected to the buses, which provides opportunities for instruction level parallelism. The parallelism is statically defined by the programmer. In this respect (and obviously due to the large instruction word width), the TTA architecture resembles the very long instruction word (VLIW) architecture. A TTA instruction word is composed of multiple slots, one slot per bus, and each slot determines the data transport that takes place on the corresponding bus. The fine-grained control allows some optimizations that are not possible in a conventional processor. For example, software can transfer data directly between functional units without using registers.

Transport triggering exposes some microarchitectural details that are normally hidden from programmers. This greatly simplifies the control logic of a processor, because many decisions normally done at run time are fixed at compile time. However, it also means that a binary compiled for one TTA processor will not run on another one without recompilation if there is even a small difference in the architecture between the two. The binary incompatibility problem, in addition to the complexity of implementing a full context switch, makes TTAs more suitable for embedded systems than for general purpose computing.

Of all the one instruction set computer architectures, the TTA architecture is one of the few that has had CPUs based on it built, and the only one that has CPUs based on it sold commercially.

Benefits in comparison to VLIW Architectures

TTAs can be seen as "exposed datapath" VLIW architectures. While VLIW is programmed using operations, TTA splits the operation execution to multiple move operations. The low level programming model enables several benefits in comparison to the standard VLIW. For example, a TTA architecture can provide more parallelism with simpler register files than with VLIW. As the programmer is in control of the timing of the operand and result data transports, the complexity (the number of input and output ports) of the register file (RF) need not be scaled according to the worst case issue/completion scenario of the multiple parallel instructions.

An important unique software optimization enabled by the transport programming is called software bypassing. In case of software bypassing, the programmer bypasses the register file write back by moving data directly to the next functional unit's operand ports. When this optimization is applied aggressively, the original move that transports the result to the register file can be eliminated completely, thus reducing both the register file port pressure and freeing a general purpose register for other temporary variables. The reduced register pressure, in addition simplifying the required complexity of the RF hardware, can lead to significant CPU energy savings, an important benefit especially in mobile embedded systems.

Structure

TTA processors are built of independent function units and register files, which are connected with transport buses and sockets.

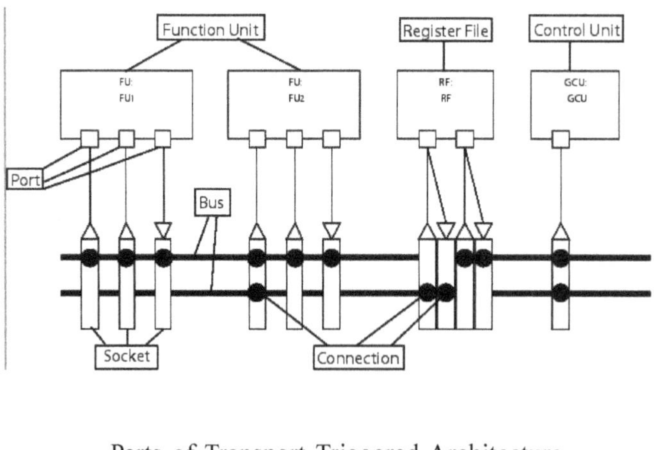

Parts of Transport Triggered Architecture

Function unit

Each function unit implements one or more operations, which implement functionality ranging from a simple addition of integers to a complex and arbitrary user-defined application-specific computation. Operands for operations are transferred through function unit *ports*.

Each function unit may have an independent pipeline. In case a function unit is fully pipelined, a new operation that takes multiple clock cycles to finish can be started in every clock cycle. On the other hand, a pipeline can be such that it does not always accept new operation start requests while an old one is still executing.

Data memory access and communication to outside of the processor is handled by using special function units. Function units that implement memory accessing operations and connect to a memory module are often called load/store units.

Control Unit

Control unit is a special case of function units which controls execution of programs. Control unit has access to the instruction memory in order to fetch the instructions to be executed. In order to allow the executed programs to transfer the execution (jump) to an arbitrary position in the executed program, control unit provides control flow operations. A control unit usually has an instruction pipeline, which consists of stages for fetching, decoding and executing program instructions.

Register Files

Register files contain general purpose registers, which are used to store variables in programs. Like function units, also register files have input and output ports. The number of read and write ports, that is, the capability of being able to read and write multiple registers in a same clock cycle, can vary in each register file.

Transport Buses and Sockets

Interconnect architecture consists of transport buses which are connected to function unit ports by means of sockets. Due to expense of connectivity, it is usual to reduce the number of connections between units (function units and register files). A TTA is said to be fully connected in case there is a path from each unit output port to every unit's input ports.

Sockets provide means for programming TTA processors by allowing to select which bus-to-port connections of the socket are enabled at any time instant. Thus, data transports taking place in a clock cycle can be programmed by defining the source and destination socket/port connection to be enabled for each bus.

Conditional Execution

Some TTA implementations support conditional execution.

Conditional execution is implemented with the aid of guards. Each data transport can be conditionalized by a guard, which is connected to a register (often a 1-bit conditional register) and to a bus. In case the value of the guarded register evaluates to false (zero), the data transport programmed for the bus the guard is connected to is squashed, that is, not written to its destination. Unconditional data transports are not connected to any guard and are always executed.

Branches

All processors, including TTA processors, include control flow instructions that alter the program counter, which are used to implement subroutines, if-then-else, for-loop, etc. The assembly language for TTA processors typically includes control flow instructions such as unconditional branches (JUMP), conditional relative branches (BNZ), subroutine call (CALL), conditional return (RETNZ), etc. that look the same as the corresponding assembly language instructions for other processors.

Like all other operations on a TTA machine, these instructions are implemented as move instructions to a special function unit.

TTA implementations that support conditional execution, such as the sTTAck and the first MOVE prototype, can implement most of these control flow instructions as a conditional move to the program counter.

TTA implementations that only support unconditional data transports, such as the MAXQ, typically have a special function unit tightly connected to the program counter that responds to a variety of destination addresses. Each such address, when used as the destination of a move , has a different effect on the program counter—each relative branch <condition> instruction has a different destination address for each condition; and other destination addresses are used CALL, RETNZ, etc.

Programming

In more traditional processor architectures, a processor is usually programmed by defining the executed operations and their operands. For example, an addition instruction in a RISC architecture could look like the following.

add r3, r1, r2

This example operation adds the values of general-purpose registers r1 and r2 and stores the result in register r3. Coarsely, the execution of the instruction in the processor probably results in translating the instruction to control signals which control the interconnection network connections and function units. The interconnection network is used to transfer the current values of registers r1 and r2 to the function unit that is capable of executing the add operation, often called ALU as in Arithmetic-Logic Unit. Finally, a control signal selects and triggers the addition operation in ALU, of which result is transferred back to the register r3.

TTA programs do not define the operations, but only the data transports needed to write and read the operand values. Operation itself is triggered by writing data to a triggering operand of an operation. Thus, an operation is executed as a side effect of the triggering data transport. Therefore, executing an addition operation in TTA requires three data transport definitions, also called moves. A move defines endpoints for a data transport taking place in a transport bus. For instance, a move can state that a data transport from function unit F, port 1, to register file R, register index 2, should take place in bus B1. In case there are multiple buses in the target processor, each bus can be utilized in parallel in the same clock cycle. Thus, it is possible to exploit data transport level parallelism by scheduling several data transports in the same instruction.

An addition operation can be executed in a TTA processor as follows:

r1 -> ALU.operand1

r2 -> ALU.add.trigger

ALU.result -> r3

The second move, a write to the second operand of the function unit called ALU, triggers the addition operation. This makes the result of addition available in the output port 'result' after the execution latency of the 'add'.

The ports associated with the ALU may act as an accumulator, allowing creation of macro instructions that abstract away the underlying TTA:

lda r1 *; "load ALU": move value to ALU operand 1*

add r2 *; add: move value to add trigger*

sta r3 *; "store ALU": move value from ALU result*

Programmer visible operation latency

The leading philosophy of TTAs is to move complexity from hardware to software. Due to this, several additional hazards are introduced to the programmer. One of them is delay slots, the programmer visible operation latency of the function units. Timing is completely the responsibility of the programmer. The programmer has to schedule the instructions such that the result is neither read too early nor too late. There is no hardware detection to lock up the processor in case a result is read too early. Consider, for example, an architecture that has an operation add with latency of 1, and operation mul with latency of 3. When triggering the add operation, it is possible to read the result in the next instruction (next clock cycle), but in case of mul, one has to wait for two instructions before the result can be read. The result is ready for the 3rd instruction after the triggering instruction.

Reading a result too early results in reading the result of a previously triggered operation, or in case no operation was triggered previously in the function unit, the read value is undefined. On the other hand, result must be read early enough to make sure the next operation result does not overwrite the yet unread result in the output port.

Due to the abundance of programmer-visible processor context which practically includes, in addition to register file contents, also function unit pipeline register contents and/or function unit input and output ports, context saves required for external interrupt support can become complex and expensive to implement in a TTA processor. Therefore, interrupts are usually not supported by TTA processors, but their task is delegated to an external hardware (e.g., an I/O processor) or their need is avoided by using an alternative synchronization/communication mechanism such as polling.

Implementations

- MAXQ

Currently, the only commercially available microcontroller built upon (though not "featuring") Transport Triggered Architecture is from Dallas Semiconductor. However, it is an OISC or "one instruction set computer", offering but a single though flexible

MOVE instruction, which can then function as various virtual instructions by moving values directly to the program counter.

- The move project has designed and fabricated several experimental TTA microprocessors.
- The TCE project is a re-implementation of the MOVE tools. The tools are available as open source, and the compiler is built around the LLVM compiler framework.
- The architecture of the Amiga Copper has all the basic features of a transport triggered architecture.
- The Able processor developed by New England Digital.
- The WireWorld based computer.
- Dr. Dobb's published One-Der a 32-bit TTA in Verilog with a matching cross assembler and Forth compiler.
- Mali (200/400) vertex processor, uses a 128-bit instruction word single precision floating point scalar TTA.

Explicitly Parallel Instruction Computing

Explicitly parallel instruction computing (EPIC) is a term coined in 1997 by the HP–Intel alliance to describe a computing paradigm that researchers had been investigating since the early 1980s. This paradigm is also called Independence architectures. It was the basis for Intel and HP development of the Intel Itanium architecture, and HP later asserted that "EPIC" was merely an old term for the Itanium architecture. EPIC permits microprocessors to execute software instructions in parallel by using the compiler, rather than complex on-die circuitry, to control parallel instruction execution. This was intended to allow simple performance scaling without resorting to higher clock frequencies.

Roots in VLIW

By 1989, researchers at HP recognized that reduced instruction set computer (RISC) architectures were reaching a limit at one instruction per cycle. They began an investigation into a new architecture, later named EPIC. The basis for the research was VLIW, in which multiple operations are encoded in every instruction, and then processed by multiple execution units.

One goal of EPIC was to move the complexity of instruction scheduling from the CPU hardware to the software compiler, which can do the instruction scheduling statically (with help of trace feedback information). This eliminates the need for complex scheduling circuitry in the CPU, which frees up space and power for other functions, including additional execution resources. An equally important goal was to further exploit instruction level parallelism (ILP), by using the compiler to find and exploit additional opportunities for parallel execution.

VLIW (at least the original forms) has several short-comings that precluded it from becoming mainstream:

- VLIW instruction sets are not backward compatible between implementations. When wider implementations (more execution units) are built, the instruction set for the wider machines is not backward compatible with older, narrower implementations.
- Load responses from a memory hierarchy which includes CPU caches and DRAM do not have a deterministic delay. This makes static scheduling of load instructions by the compiler very difficult.

EPIC architecture has evolved from VLIW architecture, while retaining many concepts of the superscalar architecture.

Moving beyond VLIW

EPIC architectures add several features to get around the deficiencies of VLIW:

- Each group of multiple software instructions is called a *bundle*. Each of the bundles has a stop bit indicating if this set of operations is depended upon by the subsequent bundle. With this capability, future implementations can be built to issue multiple bundles in parallel. The dependency information is calculated by the compiler, so the hardware does not have to perform operand dependency checking.
- A software prefetch instruction is used as a type of data prefetch. This prefetch increases the chances for a cache hit for loads, and can indicate the degree of temporal locality needed in various levels of the cache.
- A speculative load instruction is used to speculatively load data before it is known whether it will be used (bypassing control dependencies), or whether it will be modified before it is used (bypassing data dependencies).
- A check load instruction aids speculative loads by checking whether a speculative load was dependent on a later store, and thus must be reloaded.

The EPIC architecture also includes a grab-bag of architectural concepts to increase ILP:

- Predicated execution is used to decrease the occurrence of branches and to increase the speculative execution of instructions. In this feature, branch conditions are converted to predicate registers which are used to kill results of executed instructions from the side of the branch which is not taken.
- Delayed exceptions, using a not a thing bit within the general purpose registers, allow speculative execution past possible exceptions.
- Very large architectural register files avoid the need for register renaming.
- Multi-way branch instructions improve branch prediction by combining many alternative branches into one bundle.

The Itanium architecture also added rotating register files, a tool useful for software pipelining since it avoids having to manually unroll and rename registers.

Other Research and Development

There have been other investigations into EPIC architectures that are not directly tied to the development of the Itanium architecture.

- The IMPACT project at University of Illinois at Urbana-Champaign, led by Wen-mei Hwu, was the source of much influential research on this topic.
- The PlayDoh architecture from HP-labs was another major research project.
- Gelato was an open source development community in which academic and commercial researchers worked to develop more effective compilers for Linux applications running on Itanium servers.

MILL ARCHITECTURE

The Mill architecture is a novel belt machine-based computer architecture for general purpose computing. It has been under development since about 2003 by Ivan Godard and his startup Mill Computing, Inc., formerly named Out Of The Box Computing, in East Palo Alto, California. Mill Computing claims it has a "10x single-thread power/performance gain over conventional out-of-order superscalar architectures but runs the same programs, without rewrite".

Mill Computing was founded by persons who formerly worked together on a family of digital signal processors (DSPs), the Philips Trimedia.

Approach

The designers claim that the power and cost improvements come by adapting a DSP-like deeply-pipelined processor to general-purpose code. The timing hazards from branches and memory access are said to be handled via speculative execution, pipelining, and other late binding, but statically-scheduled logic. The claimed improvements in power and area are said to come from eliminating dynamic optimizing hardware: register-renaming, out-of-order execution hazard management, and dynamic cache optimizing.

The mill architecture is designed as a compiler target for highly-optimizing compilers. The overall plan is to substitute static optimizing at compile time for hardware optimizing at run time. To do this, each mill processor is designed to have timing and memory access behavior that is predictable to single cycle times.

Very Long Instruction Words

Mill uses a very long instruction word (VLIW)-style encoding to store up to 33 simple operations in wide instruction words, termed opcodes. Mill uses two program counters, and every wide instruction is split into two parts. One of the program counters counts backward. So, the code of every linear instruction block is executed from its middle to outside by two almost independent decoders. Unused operations are deleted by a small fixed-format data item in the center of each instruction. This helps maintain code density by reducing the incidence of no-operation codes in Mill code. It also allows each functional unit to start speculatively executing its instruction field, and then discard its result if it has no instruction.

The belt, a Pipelining Register System

Godard quotes research claiming that during operation of a standard register machine, values in processor registers are accessed with a frequency of: 80% once, 14% more than once, and 6% never.

Thus, the mill uses a novel temporal register addressing scheme, the belt, which has been proposed by Ivan Godard to greatly reduce the complexity of processor hardware, specifically the number of internal registers. It aids understanding to view the belt as a moving conveyor belt where the oldest values drop off the belt and vanish.

The relative-addressing aspects of the mill's machine code and assembly language may be harder to read and debug than the more conventional register name paradigm, but few large projects are written in such low-level programming languages. Eliminating registers avoids complex register renaming schemes.

A novel feature of the belt is that operations can drop multiple results to it. For example, division can produce a quotient and a remainder. Operations with overflow can produce double-wide results.

Mill instructions do not need to specify a location to store a result. Thus, they are smaller by that amount.

Godard says that the belt is not a shift register. Instead it is a semantic representation of the bypass network present in most fast computers, which intercepts pipelined accesses to registers, routing them directly to the execution units that need the result. The number of registers is reasonably small: those needed to pipeline the output of each functional unit, and one for each possible belt item. The small number of registers reduces the size, power and complexity of the network to access the registers.

Belt items are accessed by belt position, and move by changing their names in a pipeline-safe way. The names are not only belt positions, but also tags for function frames. By only incrementing the frame tag counter, the belt appears empty to a newly called function.

The length of the belt is designed so that residence time in the belt equals the time to access the scratchpad, a random-access memory (RAM) area used to spill belt items to be reused.

The belt is the fast, CPU end of a hardware caching system called the spiller, which moves belt items between subroutines, the scratchpad, static random-access memory (SRAM) buffer, and the reserved spiller memory area (backed by L2 cache) associated with each functional iteration's data area. If the bandwidth of the spiller is exceeded, the mill stalls, waiting for the belt to become consistent.

Metadata Use

Depending on the type and success of load operations, the mill also assigns metadata to each belt item, including status, width, and vectorization count. Operations operate on the item described. Thus, the width and vector count are not part of the instruction coding. If an operation fails, the failure information is hashed, and placed in the destination, with its metadata, for use in debugging.

The Mill also uses the metadata to assist speculative execution and pipelining. For example, if a vector load operation fails, e.g., part of it leaves a protection boundary, those parts of that belt entry will be marked as not a result (NaR) in the metadata. This allows speculatively-executed vector code to emulate per-vector-item fault behavior. The NaR items create a fault only if an attempt occurs to store them or perform other non-speculative code on them. If they are never used, no fault is ever created.

The mill's architecture appears able to reduce the size and complexity of pipelined loop code. It uses metadata and speculation to eliminate pipeline set-up and teardown. In the pipeline video, every operation was required to cope with an argument of not a number (NaN) in a sensible way: arithmetic and bit-wise logical operations produce a NaN if any input is a NaN. Stores and other non-speculable operations do nothing. To run a pipelined loop, the code pushes a group of NaNs on the belt, and then starts to execute the steady-state loop body. As live data iterates in the loop body, the pipeline is initialized. Teardown happens in a parallel way by feeding NaNs to the loop. A crucial invention was to allow operations to insert NaNs on the belt, for pipelined loops only.

To pipeline nested loops, the mill treats each loop almost like a subroutine call, with saves and restores of appropriate state.

Lockstep Phased Execution

Another improvement said to open up the instruction-level parallelism is that mill instructions are phased. Instructions may span several clock cycles, and hold up to 33 operations. Within an instruction, finishing occurs to math operations first, data rearrangements in the middle, and stores to memory last. Also, both the operations and even multiple cores operate in statically-predictable prioritized timings.

Family Traits

There are several versions of the mill processor in development, spanning Tin (low-end uses) to Gold (high-performance uses). The company estimates that dual-core Gold chip implemented with 28 nm lithography may work at 1.2 GHz with a typical thermal design power (TDP) of 28 Watts and performance of 79 billion operations per second.

Different versions of the mill are intended for different markets, and are said to have different instruction set architectures, different numbers of execution units, different pipeline timings, and thus, very different binaries. To accommodate these, compilers are required to emit a specification which is then recompiled into an executable binary by a recompiler supplied by the Mill Computing company. In this way, code that can be distributed is adapted to specifics of the exact model's pipeline, binary coding, etc.

The development of so many tool sets and processor designs may be impractically costly. Ivan Godard said that Mill's plan is to develop software tools that accept a specification for a mill processor, and then write the software tools (assembler, compiler backend, and simulator), and the Verilog describing the CPU. In a demo video, Mill claimed to show early versions of the software to create an assembler and simulator. The bulk of the compiler is said to be a port of LLVM. As of 2014, it is incomplete.

Criticism

In 2013, Out Of The Box Computing (OOTBC, Mill Computing) was criticized by Linley Gwennap for lack of a working compiler and of estimations in conventional benchmarks.

In the mill videos, a number of questioners said that other VLIW compiler designers have been unable to locate and use instruction-level parallelism (ILP) of more than two. Ivan Godard claims that a sufficiently-wide mill can find and use an ILP of six or more on common programs.

REFERENCES

1. Wai-Kai Chen (2000). Memory, Microprocessor, and ASIC. books.google.com. CRC Press. pp. 11–14, 11–15. Retrieved 2014-08-19.
2. Heidi Pan; Krste Asanovic (2001). Heads and Tails: A Variable-Length Instruction Format Supporting Parallel Fetch and Decode (PDF). scale.eecs.berkeley.edu. Retrieved 2014-08-19.
3. Control Data 6400/6500/6600 Computer Systems Reference Manual . 1969-02-21. Retrieved 2013-11-07.
4. Fisher, Joseph A. (1983). Very Long Instruction Word architectures and the ELI-512 (PDF). Proceedings of the 10th annual international symposium on Computer architecture. International Symposium on Computer Architecture. New York, NY, USA: Association for Computing Machinery (ACM). pp. 140–150. doi:10.1145/800046.801649. ISBN 0-89791-101-6. Retrieved 2009-04-27.
5. ACM 1985 Doctoral Dissertation Award . Association for Computing Machinery (ACM). Retrieved 2007-10-15. For his dissertation Bulldog: A Compiler for VLIW Architecture.
6. An Introduction To Very-Long Instruction Word (VLIW) Computer Architecture (PDF). Philips Semiconductors. Archived from the original (PDF) on 2011-09-29.
7. EEMBC Publishes Benchmark Scores for Infineon Technologies' Carmel® DSP Core and TriCore® *TC11IB Microcontroller"*. eembc.org. Retrieved 2016-07-28.
8. TACC . tass.ru. Retrieved 2016-07-28.
9. Pixelworks | BSP15/16 . web.archive.org. Retrieved 2016-07-28.
10. TriMedia
11. siliconhive.com - siliconhive Resources and Information. . siliconhive.com. Retrieved 2016-07-28.
12. V. Guzma, P. Jääskeläinen, P. Kellomäki, and J. Takala, "Impact of Software Bypassing on Instruction Level Parallelism and Register File Traffic"
13. Johan Janssen. "Compiler Strategies for Transport Triggered Architectures". 2001. p. 168.
14. Henk Corporaal. Transport Triggered Architectures examined for general purpose applications . p. 6.
15. Aliaksei V. Chapyzhenka. sTTAck: Stack Transport Triggered Architecture .
16. "MAXQ Family User's Guide". Section "1.1 Instruction Set" says "A register-based, transport-triggered architecture allows all instructions to be coded as

simple transfer operations. All instructions reduce to either writing an immediate value to a destination register or memory location or moving data between registers and/or memory locations.

17. Introduction to the MAXQ Architecture – Includes transfer map diagram
18. TTA Codesign Environment, an open source (MIT licensed) toolset for design of application specific TTA processors.
19. Article about TTAs, explaining how the TTA-based Codesign Environment project uses LLVM
20. Dr. Dobb's article with 32-bit FPGA CPU in Verilog
21. ETTechnology Hardware, Storage and Peripherals: Company Overview of Mill Computing, Inc. . Bloomberg. Bloomberg L.P. 23 June 2016. Retrieved 22 June 2016.
22. Kraft, Caleb (20 November 2013). The Mill: Ivan Godard Explains a Revolutionary New CPU . EETimes. A. Retrieved 23 June 2016.
23. Godard, Ivan (29 May 2013). Drinking from the firehose: How the Mill CPU decodes 30+ instructions per cycle . Stanford University Department of Electrical Engineering. Stanford University. Retrieved 23 June 2016.
24. Benchoff, Brian (2 August 2013). The Mill CPU Architecture . Hackaday. Retrieved 24 June 2016.
25. The Mill CPU Architecture, Specification (8 of 9). 2014-05-24. Retrieved 2014-07-23.
26. Gwennap, Linley (5 August 2013). Getting Way Out of the Box . Microprocessor Report: Processor Watch. The Linley Group. Retrieved 23 June 2016.
27. http://millcomputing.com/docs/belt/
28. Wai-Kai Chen (2000). Memory, Microprocessor, and ASIC. books.google.com. CRC Press. pp. 11–14, 11–15. Retrieved 2014-08-19.
29. Heidi Pan; Krste Asanovic (2001). Heads and Tails: A Variable-Length Instruction Format Supporting Parallel Fetch and Decode (PDF). scale.eecs.berkeley.edu. Retrieved 2014-08-19.
30. Fisher, Joseph A. (1983). Very Long Instruction Word architectures and the ELI-512 (PDF). Proceedings of the 10th annual international symposium on Computer architecture. International Symposium on Computer Architecture. New York, NY, USA: Association for Computing Machinery (ACM). pp. 140–150. doi:10.1145/800046.801649. ISBN 0-89791-101-6. Retrieved 2009-04-27.
31. ACM 1985 Doctoral Dissertation Award . Association for Computing Machinery (ACM). Retrieved 2007-10-15. For his dissertation Bulldog: A Compiler for VLIW Architecture.
32. An Introduction To Very-Long Instruction Word (VLIW) Computer Architecture (PDF). Philips Semiconductors. Archived from the original (PDF) on 2011-09-29.
33. EEMBC Publishes Benchmark Scores for Infineon Technologies' Carmel® DSP Core and TriCore® *TC11IB Microcontroller"*. eembc.org. Retrieved 2016-07-28.
34. TACC . tass.ru. Retrieved 2016-07-28.
35. Pixelworks | BSP15/16 . web.archive.org. Retrieved 2016-07-28.
36. TriMedia siliconhive.com - siliconhive Resources and Information. . siliconhive.com. Retrieved 2016-07-28.

CHAPTER 9

Finite Impulse Response

Chapter Contents

Finite impulse response (FIR) filters are the most popular type of filters implemented in software. This introduction will help you understand them both on a theoretical and a practical level.

Filters are signal conditioners. Each functions by accepting an input signal, blocking prespecified frequency components, and passing the original signal minus those components to the output. For example, a typical phone line acts as a filter that limits frequencies to a range considerably smaller than the range of frequencies human beings can hear. That's why listening to CD-quality music over the phone is not as pleasing to the ear as listening to it directly.

INTRODUCTION

In signal processing, a finite impulse response (FIR) filter is a filter whose impulse response (or response to any finite length input) is of finite duration, because it settles to zero in finite time. This is in contrast to infinite impulse response (IIR) filters, which may have internal feedback and may continue to respond indefinitely (usually decaying).

The impulse response (that is, the output in response to a Kronecker delta input) of an Nth-order discrete-time FIR filter lasts exactly N + 1 samples (from first nonzero element through last nonzero element) before it then settles to zero.

FIR filters can be discrete-time or continuous-time, and digital or analog.

DEFINITION

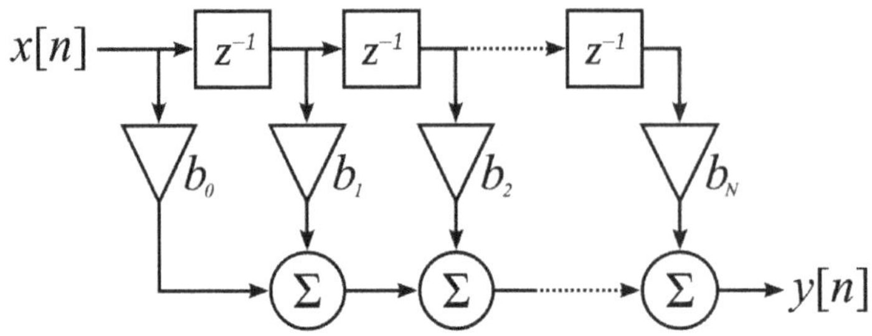

A direct form discrete-time FIR filter of order N. The top part is an N-stage delay line with $N + 1$ taps. Each unit delay is a z^{-1} operator in Z-transform notation.

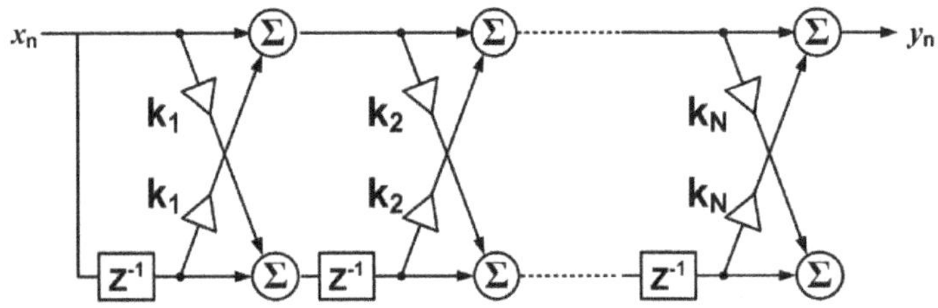

A lattice-form discrete-time FIR filter of order N. Each unit delay is a z^{-1} operator in Z-transform notation.

For a causal discrete-time FIR filter of order N, each value of the output sequence is a weighted sum of the most recent input values:

$$\begin{aligned} y[n] &= b_0 x[n] + b_1 x[n-1] + \cdots + b_N x[n-N] \\ &= \sum_{i=0}^{N} b_i \cdot x[n-i], \end{aligned}$$

where:

- x[n]is the input signal,
- y[n] is the output signal,

- N is the filter order; an Nth-order filter has (N+1) terms on the right-hand side
- $x[n-i]$ is the value of the impulse response at the i'th instant for $0 \le i \le N$ of an Nth-order FIR filter. If the filter is a direct form FIR filter then b_i is also a coefficient of the filter.

This computation is also known as discrete convolution.

The $x[n-i]$ in these terms are commonly referred to as taps, based on the structure of a tapped delay line that in many implementations or block diagrams provides the delayed inputs to the multiplication operations. One may speak of a 5th order/6-tap filter, for instance.

The impulse response of the filter as defined is nonzero over a finite duration. Including zeros, the impulse response is the infinite sequence:

$$h[n] = \sum_{i=0}^{N} b_i \cdot \delta[n-i] = \begin{cases} b_n & 0 \le n \le N \\ 0 & \text{otherwise.} \end{cases}$$

If an FIR filter is non-causal, the range of nonzero values in its impulse response can start before n = 0, with the defining formula appropriately generalized.

PROPERTIES

An FIR filter has a number of useful properties which sometimes make it preferable to an infinite impulse response (IIR) filter. FIR filters:

- Require no feedback. This means that any rounding errors are not compounded by summed iterations. The same relative error occurs in each calculation. This also makes implementation simpler.
- Are inherently stable, since the output is a sum of a finite number of finite multiples of the input values, so can be no greater than $\sum |b_i|$ times the largest value appearing in the input.
- Can easily be designed to be linear phase by making the coefficient sequence symmetric. This property is sometimes desired for phase-sensitive applications, for example data communications, seismology, crossover filters, and mastering.

The main disadvantage of FIR filters is that considerably more computation power in a general purpose processor is required compared to an IIR filter with similar sharpness or selectivity, especially when low frequency (relative to the sample rate) cutoffs are needed. However many digital signal processors provide specialized hardware features to make FIR filters approximately as efficient as IIR for many applications.

FREQUENCY RESPONSE

The filter's effect on the sequence x[n] is described in the frequency domain by the convolution theorem:

$$\underbrace{\mathcal{F}\{x * h\}}_{Y(\omega)} = \underbrace{\mathcal{F}\{x\}}_{X(\omega)} \cdot \underbrace{\mathcal{F}\{h\}}_{H(\omega)} \quad \text{and} \quad y[n] = x[n] * h[n] = \mathcal{F}^{-1}\big\{X(\omega) \cdot H(\omega)\big\},$$

where operators $\mathcal{F}$ and $\mathcal{F}^{-1}$ respectively denote the discrete-time Fourier transform (DTFT) and its inverse. Therefore, the complex-valued, multiplicative function $H(\omega)$ is the filter's frequency response. It is defined by a Fourier series:

$$H_{2\pi}(\omega) \stackrel{\text{def}}{=} \sum_{n=-\infty}^{\infty} h[n] \cdot \left(e^{i\omega}\right)^{-n} = \sum_{n=0}^{N} b_n \cdot \left(e^{i\omega}\right)^{-n},$$

where the added subscript denotes 2π-periodicity. Here ω represents frequency in normalized units (radians/sample). The substitution $\omega = 2\pi f,$ favored by many filter design programs, changes the units of frequency f to cycles/sample and the periodicity to 1.[note 1] When the x[n] sequence has a known sampling-rate, f_s samples/second, the substitution $\omega = 2\pi f/f_s$ changes the units of frequency f_s to cycles/second (hertz) and the periodicity to f_s The value $\omega = \pi$ corresponds to a frequency of $f = \frac{f_s}{2}\,Hz = \frac{1}{2}$ cycles/sample, which is the Nyquist frequency.

TRANSFER FUNCTION

The frequency response $H_{2\pi}(\omega)$, can also be written as $H(e^{i\omega})$, where function H is the Z- transform of the impulse response:

$$H(z) \stackrel{\text{def}}{=} \sum_{n=-\infty}^{\infty} h[n] \cdot z^{-n}.$$

z is a complex variable, and H(z) is a surface. One cycle of the periodic frequency response can be found in the region defined by $z = e^{i\omega}$, $-\pi \le \omega \le \pi,$ which is the unit circle of the z-plane. Filter transfer functions are often used to verify the stability of IIR designs. As we have already noted, FIR designs are inherently stable.

Filter Design

An FIR filter is designed by finding the coefficients and filter order that meet certain specifications, which can be in the time-domain (e.g. a matched filter) and/or the frequency domain (most common). Matched filters perform a cross-correlation between the input signal and a known pulse-shape. The FIR convolution is a cross-correlation between the input signal and a time-reversed copy of the impulse-response. Therefore, the matched-filter's impulse response is "designed" by sampling the known pulse-shape and using those samples in reverse order as the coefficients of the filter.

When a particular frequency response is desired, several different design methods are common:

- Window design method
- Frequency Sampling method
- Weighted least squares design
- Parks-McClellan method (also known as the Equiripple, Optimal, or Minimax method). The Remez exchange algorithm is commonly used to find an optimal equiripple set of coefficients. Here the user specifies a desired frequency response, a weighting function for errors from this response, and a filter order *N*. The algorithm then finds

the set of N+1 coefficients that minimize the maximum deviation from the ideal. Intuitively, this finds the filter that is as close as you can get to the desired response given that you can use only N+1 coefficients. This method is particularly easy in practice since at least one text includes a program that takes the desired filter and *N*, and returns the optimum coefficients.

- Equiripple FIR filters can be designed using the FFT algorithms as well. The algorithm is iterative in nature. You simply compute the DFT of an initial filter design that you have using the FFT algorithm (if you don't have an initial estimate you can start with h[n]=delta[n]). In the Fourier domain or FFT domain you correct the frequency response according to your desired specs and compute the inverse FFT. In time-domain you retain only N of the coefficients (force the other coefficients to zero). Compute the FFT once again. Correct the frequency response according to specs.

Software packages like MATLAB, GNU Octave, Scilab, and SciPy provide convenient ways to apply these different methods.

Window Design Method

In the window design method, one first designs an ideal IIR filter and then truncates the infinite impulse response by multiplying it with a finite length window function. The result is a finite impulse response filter whose frequency response is modified from that of the IIR filter. Multiplying the infinite impulse by the window function in the time domain results in the frequency response of the IIR being convolved with the Fourier transform (or DTFT) of the window function. If the window's main lobe is narrow, the composite frequency response remains close to that of the ideal IIR filter.

The ideal response is usually rectangular, and the corresponding IIR is a sinc function. The result of the frequency domain convolution is that the edges of the rectangle are tapered, and ripples appear in the passband and stopband. Working backward, one can specify the slope (or width) of the tapered region (*transition band*) and the height of the ripples, and thereby derive the frequency domain parameters of an appropriate window function. Continuing backward to an impulse response can be done by iterating a filter design program to find the minimum filter order. Another method is to restrict the solution set to the parametric family of Kaiser windows, which provides closed form relationships between the time-domain and frequency domain parameters. In general, that method will not achieve the minimum possible filter order, but it is particularly convenient for automated applications that require dynamic, on-the-fly, filter design.

The window design method is also advantageous for creating efficient half-band filters, because the corresponding sinc function is zero at every other sample point (except the center one). The product with the window function does not alter the zeros, so almost half of the coefficients of the final impulse response are zero. An appropriate implementation of the FIR calculations can exploit that property to double the filter's efficiency.

Moving Average Example

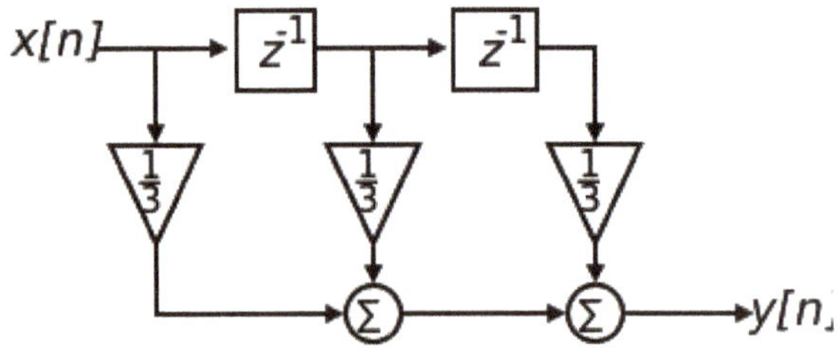

Figure. (a) Block diagram of a simple FIR filter (2nd-order/3-tap filter in this case, implementing a moving average)

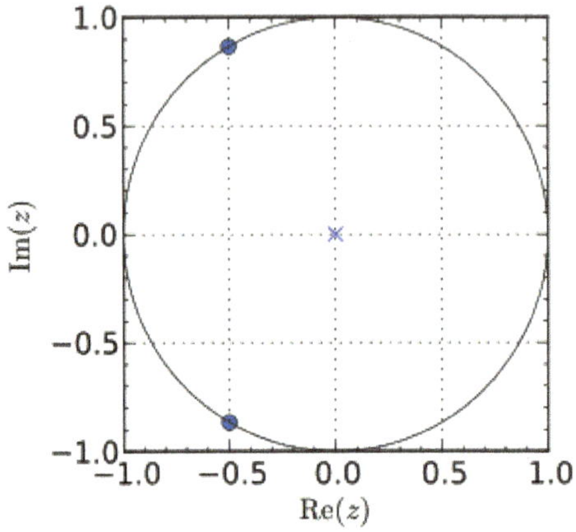

Figure. (b) Pole–zero diagram of a second-order FIR filter

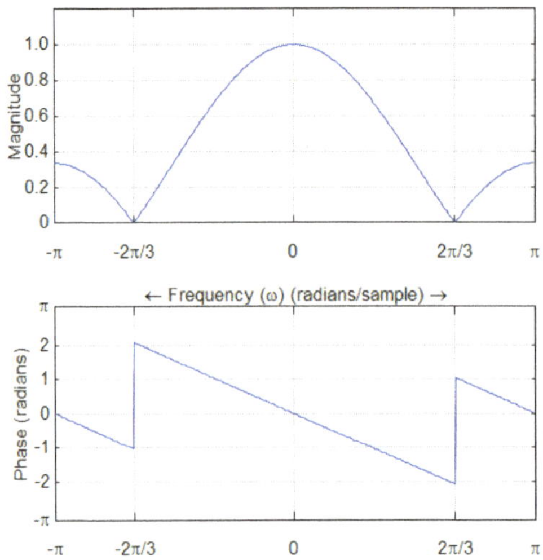

Figure. (c) Magnitude and phase responses

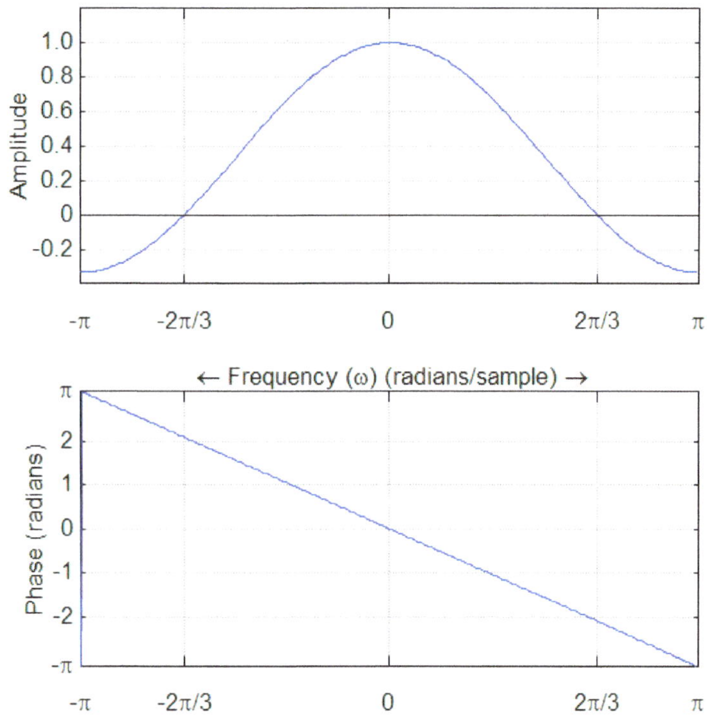

Figure. (d) Amplitude and phase responses

A moving average filter is a very simple FIR filter. It is sometimes called a boxcar filter, especially when followed by decimation. The filter coefficients, $b_0, \ldots, b_N$, are found via the following equation:

$$b_i = \frac{1}{N+1}$$

To provide a more specific example, we select the filter order:

$$N = 2$$

The impulse response of the resulting filte

r is:

$$h[n] = \frac{1}{3}\delta[n] + \frac{1}{3}\delta[n-1] + \frac{1}{3}\delta[n-2]$$

The Fig. (a) on the right shows the block diagram of a 2nd-order moving-average filter discussed below. The transfer function is:

$$H(z) = \frac{1}{3} + \frac{1}{3}z^{-1} + \frac{1}{3}z^{-2} = \frac{1}{3}\frac{z^2+z+1}{z^2}.$$

Fig. (b) on the right shows the corresponding pole–zero diagram. Zero frequency (DC) corresponds to (1,0), positive frequencies advancing counterclockwise around the circle to the Nyquist frequency at (-1,0). Two poles are located at the origin, and two zeros

$$z_1 = -\frac{1}{2} + j\frac{\sqrt{3}}{2},\ z_2 = -\frac{1}{2} - j\frac{\sqrt{3}}{2}.$$

are located at

The frequency response, in terms of normalized frequency ω, is:

$$H\left(e^{j\omega}\right) = \frac{1}{3} + \frac{1}{3}e^{-j\omega} + \frac{1}{3}e^{-j2\omega}.$$

Fig. (c) on the right shows the magnitude and phase components of $H(e^{j\omega})$ But plots like these can also be generated by doing a discrete Fourier transform (DFT) of the impulse response.[note 2] And because of symmetry, filter design or viewing software often displays only the [0,π] region. The magnitude plot indicates that the moving-average filter passes low frequencies with a gain near 1 and attenuates high frequencies, and is thus a crude low-pass filter. The phase plot is linear except for discontinuities at the two frequencies where the magnitude goes to zero. The size of the discontinuities is π, representing a sign reversal. They do not affect the property of linear phase. That fact is illustrated in Fig. (d).

FILTER DESIGN

Filter design is the process of designing a signal processing filter that satisfies a set of requirements, some of which are contradictory. The purpose is to find a realization of the filter that meets each of the requirements to a sufficient degree to make it useful.

The filter design process can be described as an optimization problem where each requirement contributes to an error function which should be minimized. Certain parts of the design process can be automated, but normally an experienced electrical engineer is needed to get a good result.

Typical Design Requirements

Typical requirements which are considered in the design process are:

- The filter should have a specific frequency response
- The filter should have a specific phase shift or group delay
- The filter should have a specific impulse response
- The filter should be causal
- The filter should be stable
- The filter should be localized (pulse or step inputs should result in finite time outputs)
- The computational complexity of the filter should be low
- The filter should be implemented in particular hardware or software

The Frequency Function

An important parameter is the required frequency response. In particular, the steepness and complexity of the response curve is a deciding factor for the filter order and feasibility.

A first order recursive filter will only have a single frequency-dependent component. This means that the slope of the frequency response is limited to 6 dB per octave. For many purposes, this is not sufficient. To achieve steeper slopes, higher order filters are required.

In relation to the desired frequency function, there may also be an accompanying weighting function which describes, for each frequency, how important it is that the resulting frequency function approximates the desired one. The larger weight, the more important is a close approximation.

Typical examples of frequency function are:

- A low-pass filter is used to cut unwanted high-frequency signals.
- A high-pass filter passes high frequencies fairly well; it is helpful as a filter to cut any unwanted low frequency components.
- A band-pass filter passes a limited range of frequencies.
- A band-stop filter passes frequencies above and below a certain range. A very narrow band-stop filter is known as a notch filter.
- A differentiator has an amplitude response proportional to the frequency.
- A low-shelf filter passes all frequencies, but increases or reduces frequencies below the shelf frequency by specified amount.
- A high-shelf filter passes all frequencies, but increases or reduces frequencies above the shelf frequency by specified amount.
- A peak EQ filter makes a peak or a dip in the frequency response, commonly used in parametric equalizers.

Phase and Group Delay

- An all-pass filter passes through all frequencies unchanged, but changes the phase of the signal. Filters of this type can be used to equalize the group delay of recursive filters. This filter is also used in phaser effects.
- A Hilbert transformer is a specific all-pass filter that passes sinusoids with unchanged amplitude but shifts each sinusoid phase by ±90°.
- A fractional delay filter is an all-pass that has a specified and constant group or phase delay for all frequencies.

The Impulse Response

There is a direct correspondence between the filter's frequency function and its impulse response: the former is the Fourier transform of the latter. That means that any requirement on the frequency function is a requirement on the impulse response, and vice versa.

However, in certain applications it may be the filter's impulse response that is explicit and the design process then aims at producing as close an approximation as possible to the requested impulse response given all other requirements.

In some cases it may even be relevant to consider a frequency function and impulse response of the filter which are chosen independently from each other. For example, we may want both a specific frequency function of the filter and that the resulting filter have a small effective width in the signal domain as possible. The latter condition can be realized by considering a very narrow function as the wanted impulse response of the filter even though this function has no relation to the desired frequency function. The goal of the design process is then to realize a filter which tries to meet both these contradicting design goals as much as possible.

Causality

In order to be implementable, any time-dependent filter (operating in real time) must be causal: the filter response only depends on the current and past inputs. A standard approach is to leave this requirement until the final step. If the resulting filter is not causal, it can be made causal by introducing an appropriate time-shift (or delay). If the filter is a part of a larger system (which it normally is) these types of delays have to be introduced with care since they affect the operation of the entire system.

Filters that do not operate in real time (e.g. for image processing) can be non-causal. This e.g. allows the design of zero delay recursive filters, where the group delay of a causal filter is canceled by its Hermitian non-causal filter.

Stability

A stable filter assures that every limited input signal produces a limited filter response. A filter which does not meet this requirement may in some situations prove useless or even harmful. Certain design approaches can guarantee stability, for example by using only feed-forward circuits such as an FIR filter. On the other hand, filters based on feedback circuits have other advantages and may therefore be preferred, even if this class of filters includes unstable filters. In this case, the filters must be carefully designed in order to avoid instability.

Locality

In certain applications we have to deal with signals which contain components which can be described as local phenomena, for example pulses or steps, which have certain time duration. A consequence of applying a filter to a signal is, in intuitive terms, that the duration of the local phenomena is extended by the width of the filter. This implies that it is sometimes important to keep the width of the filter's impulse response function as short as possible.

According to the uncertainty relation of the Fourier transform, the product of the width of the filter's impulse response function and the width of its frequency function must exceed a certain constant. This means that any requirement on the filter's locality also implies a bound on its frequency function's width. Consequently, it may not be possible to simultaneously meet requirements on the locality of the filter's impulse response function as well as on its frequency function. This is a typical example of contradicting requirements.

Computational Complexity

A general desire in any design is that the number of operations (additions and multiplications) needed to compute the filter response is as low as possible. In certain applications, this desire is a strict requirement, for example due to limited computational resources, limited power resources, or limited time. The last limitation is typical in real-time applications.

There are several ways in which a filter can have different computational complexity. For example, the order of a filter is more or less proportional to the number of operations. This means that by choosing a low order filter, the computation time can be reduced.

For discrete filters the computational complexity is more or less proportional to the number of filter coefficients. If the filter has many coefficients, for example in the case of multidimensional signals such as tomography data, it may be relevant to reduce the number of coefficients by removing those which are sufficiently close to zero. In multirate filters, the number of coefficients by taking advantage of its bandwidth limits, where the input signal is downsampled (e.g. to its critical frequency), and upsampled after filtering.

Another issue related to computational complexity is separability, that is, if and how a filter can be written as a convolution of two or more simpler filters. In particular, this issue is of importance for multidimensional filters, e.g., 2D filter which are used in image processing. In this case, a significant reduction in computational complexity can be obtained if the filter can be separated as the convolution of one 1D filter in the horizontal direction and one 1D filter in the vertical direction. A result of the filter design process may, e.g., be to approximate some desired filter as a separable filter or as a sum of separable filters.

Other Considerations

It must also be decided how the filter is going to be implemented:

- Analog filter
- Analog sampled filter
- Digital filter
- Mechanical filter

Analog Filters

The design of linear analog filters is for the most part covered in the linear filter section.

Digital Filters

Digital filters are classified into one of two basic forms, according to how they respond to a unit impulse:

- Finite impulse response, or **FIR**, filters express each output sample as a weighted sum of the last *N* input samples, where *N* is the order of the filter. FIR filters are normally non-recursive, meaning they do not use feedback and as such are inherently stable. A moving average filter or CIC filter are examples of FIR filters that are normally recursive (that use feedback). If the FIR coefficients are symmetrical (often the case), then such a filter is linear phase, so it delays signals of all frequencies equally which is important in many applications. It is also straightforward to avoid overflow in an FIR filter. The main disadvantage is that they may require significantly more processing and memory resources than cleverly designed IIR variants. FIR filters are generally easier to design than IIR filters - the Parks-McClellan filter design algorithm (based on the Remez algorithm) is one suitable method for designing quite good filters semi-automatically. (See Methodology.)
- Infinite impulse response, or **IIR**, filters are the digital counterpart to analog filters. Such a filter contains internal state, and the output and the next internal state are determined by a linear combination of the previous inputs and outputs (in other words, they use feedback, which FIR filters normally do not). In theory, the impulse response of such a filter never dies out completely, hence the name IIR, though in practice, this is not true given the finite resolution of computer arithmetic. IIR filters normally require less computing resources than an FIR filter of similar performance. However, due to the feedback, high order IIR filters may have problems with instability, arithmetic overflow, and limit cycles, and require careful design to avoid such pitfalls. Additionally, since the phase shift is inherently a non-linear function of frequency, the time delay through such a filter is frequency-dependent, which can be a problem in many situations. 2nd order IIR filters are often called 'biquads' and a common implementation of higher order filters is to cascade biquads. A useful reference for computing biquad coefficients is the RBJ Audio EQ Cookbook.

Sample rate

Unless the sample rate is fixed by some outside constraint, selecting a suitable sample rate is an important design decision. A high rate will require more in terms of computational resources, but less in terms of anti-aliasing filters. Interference and beating with other signals in the system may also be an issue.

Anti-aliasing

For any digital filter design, it is crucial to analyze and avoid aliasing effects. Often, this is done by adding analog anti-aliasing filters at the input and output, thus avoiding any frequency component above the Nyquist frequency. The complexity (i.e., steepness) of such filters depends on the required signal to noise ratio and the ratio between the sampling rate and the highest frequency of the signal.

Theoretical Basis

Parts of the design problem relate to the fact that certain requirements are described in the frequency domain while others are expressed in the signal domain and that these may contradict. For example, it is not possible to obtain a filter which has both an arbitrary impulse response and arbitrary frequency function. Other effects which refer to relations between the signal and frequency domain are

- The uncertainty principle between the signal and frequency domains
- The variance extension theorem
- The asymptotic behaviour of one domain versus discontinuities in the other

The Uncertainty Principle

As stated in the uncertainty principle, the product of the width of the frequency function and the width of the impulse response cannot be smaller than a specific constant. This implies that if a specific frequency function is requested, corresponding to a specific frequency width, the minimum width of the filter in the signal domain is set. Vice versa, if the maximum width of the response is given, this determines the smallest possible width in the frequency. This is a typical example of contradictory requirements where the filter design process may try to find a useful compromise.

The variance extension theorem

Let σ_s^2 be the variance of the input signal and let σ_f^2 be the variance of the filter. The variance of the filter response, σ_r^2, is then given by

$$\sigma_r^2 = \sigma_s^2 + \sigma_f^2$$

This means that $\sigma_r > \sigma_f$ and implies that the localization of various features such as pulses or steps in the filter response is limited by the filter width in the signal domain. If a precise localization is requested, we need a filter of small width in the signal domain and, via the uncertainty principle, its width in the frequency domain cannot be arbitrary small.

Discontinuities Versus Asymptotic Behaviour

Let f(t) be a function and let $F(\omega)$ be its Fourier transform. There is a theorem which states that if the first derivative of F which is discontinuous has order $n \geq 0$, then f has an asymptotic decay like t^{-n-1}.

A consequence of this theorem is that the frequency function of a filter should be as smooth as possible to allow its impulse response to have a fast decay, and thereby a short width.

Methodology

One common method for designing FIR filters is the Parks-McClellan filter design algorithm, based on the Remez exchange algorithm. Here the user specifies a desired frequency response, a weighting function for errors from this response, and a filter order N. The algorithm then finds the set of N coefficients that minimize the maximum deviation from the ideal. Intuitively, this finds the filter that is as close as you can get to the desired response given that you can use only N coefficients. This method is particularly easy in practice and at least one text includes a program that takes the desired filter and N and returns the optimum coefficients. One possible drawback to filters designed this way is that they contain many small ripples in the passband(s), since such a filter minimizes the peak error.

Another method to finding a discrete FIR filter is filter optimization described in Knutsson et al., which minimizes the integral of the square of the error, instead of its maximum value. In its basic form this approach requires that an ideal frequency function of the filter $F_I(\omega)$ is specified together with a frequency weighting function $W(\omega)$ in the signal domain where the filter coefficients are located.

An error function ε is defined as

$$\varepsilon = \|W \cdot (F_I - \mathcal{F}\{f\})\|^2$$

where $f(x)$ is the discrete filter and $\mathcal{F}$ is the discrete-time Fourier transform defined on the specified set of coordinates. The norm used here is, formally, the usual norm on L^2 spaces. This means that ε measures the deviation between the requested frequency function of the filter, F_I, and the actual frequency function of the realized filter, $\mathcal{F}\{f\}$. However, the deviation is also subject to the weighting function W before the error function is computed.

Once the error function is established, the optimal filter is given by the coefficients $f(x)$ which minimize ε. This can be done by solving the corresponding least squares problem. In practice, the L^2 norm has to be approximated by means of a suitable sum over discrete points in the frequency domain. In general, however, these points should be significantly more than the number of coefficients in the signal domain to obtain a useful approximation.

Simultaneous Optimization in Both Domains

The previous method can be extended to include an additional error term related to a desired filter impulse response in the signal domain, with a corresponding weighting function. The ideal impulse response can be chosen independently of the ideal frequency function and is in practice used to limit the effective width and to remove ringing effects of the resulting filter in the signal domain. This is done by choosing a narrow ideal filter impulse response function, e.g., an impulse, and a weighting function which grows fast with the distance from the origin, e.g., the distance squared. The optimal filter can still be calculated by solving a simple least squares problem and the resulting filter is then a "compromise" which has a total optimal fit to the ideal functions in both domains. An important parameter is the relative strength of the two weighting functions which determines in which domain it is more important to have a good fit relative to the ideal function.

REFERENCES

1. Antoniou (1993). Digital Filters: Analysis, Design, and Applications (2 ed.). McGraw-Hill, New York, NY. ISBN 0-07-002117-1.
2. Antoniou (2006). Digital Signal Processing: Signals, Systems, and Filters. McGraw-Hill, New York, NY. doi:10.1036/0071454241. ISBN 0-07-145424-1.
3. S.W.A. Bergen; A. Antoniou (2005). "Design of Nonrecursive Digital Filters Using the Ultraspherical Window Function". EURASIP Journal on Applied Signal Processing. **2005** *(12): 1910. doi:10.1155/ASP.2005.1910.*
4. A.G. Deczky (October 1972). Synthesis of Recursive Digital Filters Using the Minimum p-Error Criterion. IEEE Trans. Audio Electroacoust. AU-20 (4): 257–263. doi:10.1109/TAU.1972.1162392.
5. J.K. Kaiser (1974). Nonrecursive Digital Filter Design Using the I_0-sinh Window Function". Proc. 1974 IEEE Int. Symp. Circuit Theory (ISCAS74). San Francisco, CA. pp. 20–23.
6. H. Knutsson; M. Andersson; J. Wiklund (June 1999). "Advanced Filter Design". Proc. Scandinavian Symposium on Image Analysis, Kangerlussuaq, Greenland.
7. S.K. Mitra (1998). Digital Signal Processing: A Computer-Based Approach. McGraw-Hill, New York, NY. ISBN 0-07-286546-6.
8. A.V. Oppenheim; R.W. Schafer; J.R. Buck (1999). Discrete-Time Signal Processing. Prentice-Hall, Upper Saddle River, NJ. ISBN 0-13-754920-2.
9. T.W. Parks; J.H. McClellan (March 1972). "Chebyshev Approximation for Nonrecursive Digital Filters with Linear Phase". IEEE Trans. Circuit Theory. CT-19 (2): 189–194. doi:10.1109/TCT.1972.1083419.
10. L.R. Rabiner; J.H. McClellan; T.W. Parks (April 1975). "FIR Digital Filter Design Techniques Using Weighted Chebyshev Approximation". Proc. IEEE. **63** *(4): 595–610. doi:10.1109/PROC.1975.9794.*
11. F. Alton Everest (2000). Master Handbook of Acoustics (Fourth ed.). McGraw-Hill Professional. ISBN 0-07-136097-2.
12. Monitor: Proceedings of the Institution of Radio and Electronics Engineers Australia (1976)
13. http://www.acoustics.hut.fi/projects/poririrs/ the Concert Hall Impulse Responses from Pori, Finland
14. Lütkepohl, Helmut (2008). Impulse response function. The New Palgrave Dictionary of Economics (2nd ed.).
15. Hamilton, James D. (1994). Difference Equations. Time Series Analysis. Princeton University Press. p. 5. ISBN 0-691-04289-6.
16. Hatemi-J, A. (2014). Asymmetric generalized impulse responses with an application in finance. Economic Modelling. **36**: 22–24. doi:10.1016/j.econmod.2013.09.014.
17. Oppenheim, Alan V., Willsky, Alan S., and Young, Ian T.,1983: Signals and Systems, p. 256 (Englewood Cliffs, New Jersey: Prentice-Hall, Inc.) ISBN 0-13-809731-3
18. Rabiner, Lawrence R., and Gold, Bernard, 1975: Theory and Application of Digital Signal Processing (Englewood Cliffs, New Jersey: Prentice-Hall, Inc.) ISBN 0-13-914101-4
19. E. Cetin, O.N. Gerek, Y. Yardimci, Equiripple FIR filter design by the FFT algorithm, IEEE Signal Processing Magazine, pp. 60-64, March 1997.

CHAPTER 10

Convolution Applications to DSP

Chapter Contents

- **Introduction**
- **Definition**
- **Visual Explanation**
- **Historical Developments**
- **Circular Convolution**
- **Circular Discrete Convolution**
- **Properties**
- **Bialgebras**
- **Applications**

Recall that a grayscale intensity image can be represented as a matrix of numbers, where the numbers are called pixel values. Computational methods in image processing use mathematical operations on these pixel values to manipulate (e.g., rotate, translate, zoom in/out, enhance, restore) images. For example, we could brighten certain areas of an image by in- creasing the intensity of the corresponding pixel values. In this set of notes we consider an important image processing computation that occurs in many applications: convolution.

INTRODUCTION

In mathematics (and, in particular, functional analysis) convolution is a mathematical operation on two functions (f and g); it produces a third function, that is typically viewed as a modified version of one of the original functions, giving the integral of the pointwise multiplication of the two functions as a function of the amount that one of the original functions is translated. Convolution is similar to cross-correlation. It has applications that include probability, statistics, computer vision, natural language processing, image and signal processing, engineering, and differential equations.

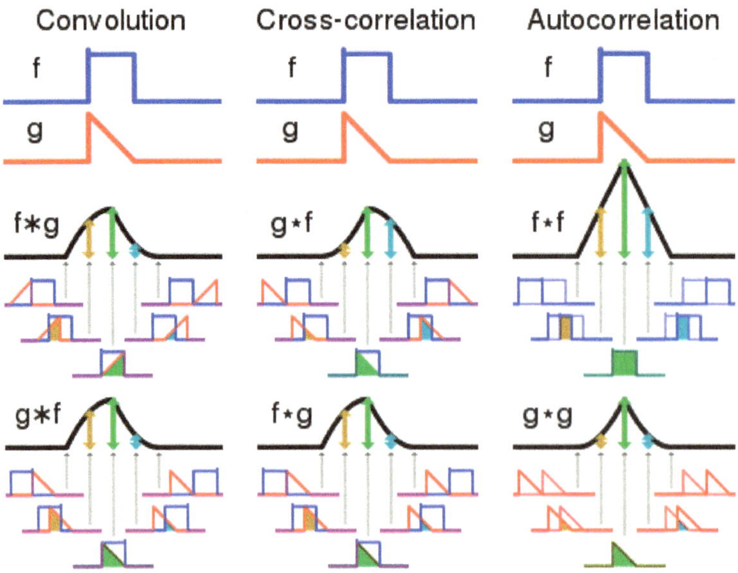

Convolution and correlation. Note that the symmetry of function f causes $g * f$ and $f \star g$ to be identical operations with identical results.

The convolution can be defined for functions on groups other than Euclidean space. For example, periodic functions, such as the discrete-time Fourier transform, can be defined on a circle and convolved by periodic convolution. (See row 10 at DTFT#Properties.) A discrete convolution can be defined for functions on the set of integers. Generalizations of convolution have applications in the field of numerical analysis and numerical linear algebra, and in the design and implementation of finite impulse response filters in signal processing.

Computing the inverse of the convolution operation is known as deconvolution.

DEFINITION

The convolution of f and g is written f∗g, using an asterisk or star. It is defined as the integral of the product of the two functions after one is reversed and shifted. As such, it is a particular kind of integral transform:

$$(f * g)(t) \stackrel{\text{def}}{=} \int_{-\infty}^{\infty} f(\tau)\, g(t-\tau)\, d\tau$$
$$= \int_{-\infty}^{\infty} f(t-\tau)\, g(\tau)\, d\tau.$$

While the symbol t is used above, it need not represent the time domain. But in that context, the convolution formula can be described as a weighted average of the function

$f(\tau)$ at the moment t where the weighting is given by $g(-\tau)$ simply shifted by amount t. As t changes, the weighting function emphasizes different parts of the input function. For functions f, g supported on only $[0, \infty)$ (i.e., zero for negative arguments), the integration limits can be truncated, resulting in

$$(f * g)(t) = \int_0^t f(\tau)\, g(t - \tau)\, d\tau \text{ for } f, g : [0, \infty) \to \mathbb{R}$$

In this case, the Laplace transform is more appropriate than the Fourier transform below and boundary terms become relevant.

For the multi-dimensional formulation of convolution, see Domain of definition (below).

Notation

A primarily engineering convention that one often sees is:

$$f(t) * g(t) \stackrel{\text{def}}{=} \underbrace{\int_{-\infty}^{\infty} f(\tau)\, g(t - \tau)\, d\tau}_{(f*g)(t)},$$

which has to be interpreted carefully to avoid confusion. For instance, $f(t)*g(t - t_0)$ is equivalent to $(f*g)(t - t_0)$, but $f(t - t_0)*g(t - t_0)$ is in fact equivalent to $(f*g)(t-2t_0)$.

Derivations

Convolution describes the output (in terms of the input) of an important class of operations known as linear time-invariant (LTI). See LTI system theory for a derivation of convolution as the result of LTI constraints. In terms of the Fourier transforms of the input and output of an LTI operation, no new frequency components are created. The existing ones are only modified (amplitude and/or phase). In other words, the output transform is the pointwise product of the input transform with a third transform (known as a transfer function). See Convolution theorem for a derivation of that property of convolution. Conversely, convolution can be derived as the inverse Fourier transform of the pointwise product of two Fourier transforms.

VISUAL EXPLANATION

Visual explanations of convolution

1. Express each function in terms of a dummy variable τ. 2. Reflect one of the functions: $g(\tau) \rightarrow g(-\tau)$. 3. Add a time-offset, t, which allows $g(t-\tau)$ to slide along the τ-axis. 4. Start t at $-\infty$ and slide it all the way to $+\infty$. Wherever the two functions intersect, find the integral of their product. In other words, compute a sliding, weighted-sum of function $f(\tau)$, where the weighting function is $g(-\tau)$. The resulting waveform (not shown here) is the convolution of functions f and g. If $f(t)$ is a unit impulse, the result of this process is simply $g(t)$, which is therefore called the impulse response. Formally: $\int_{-\infty}^{\infty} \delta(\tau)\, g(t-\tau)\, d\tau = g(t)$	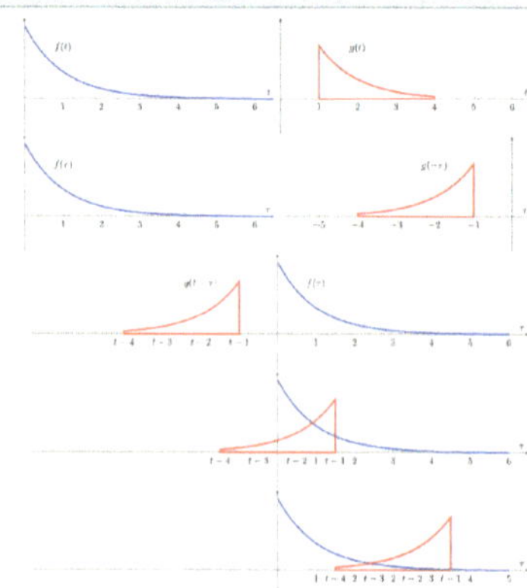
In this example, the red-colored "pulse", $g(\tau)$, is a symmetrical function ($g(-\tau) = g(\tau)$), so convolution is equivalent to correlation. A snapshot of this "movie" shows functions $g(t-\tau)$ and $f(\tau)$ (in blue) for some value of parameter t, which is arbitrarily defined as the distance from the $\tau = 0$ axis to the center of the red pulse. The amount of yellow is the area of the product $f(\tau) \cdot g(t-\tau)$, computed by the convolution/correlation integral. The movie is created by continuously changing t and recomputing the integral. The result (shown in black) is a function of t, but is plotted on the same axis as τ, for convenience and comparison.	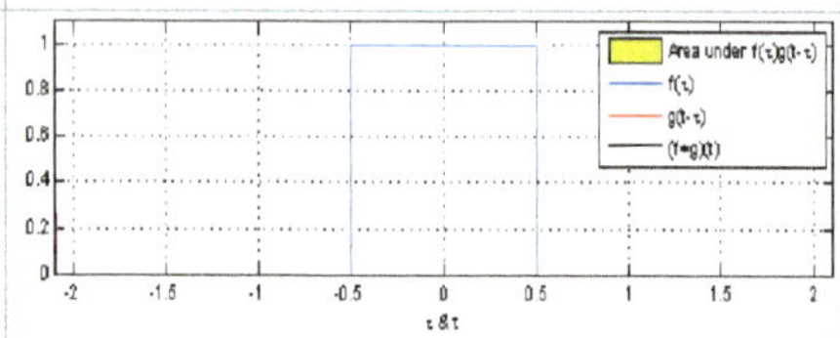
In this depiction, $f(\tau)$ could represent the response of an RC circuit to a narrow pulse that occurs at $\tau = 0$. In other words, if $g(\tau) = \delta(\tau)$, the result of convolution is just $f(t)$. But when $g(\tau)$ is the wider pulse (in red), the response is a "smeared" version of $f(t)$. It begins at $t = -0.5$, because we defined t as the distance from the $\tau = 0$ axis to the center of the wide pulse (instead of the leading edge).	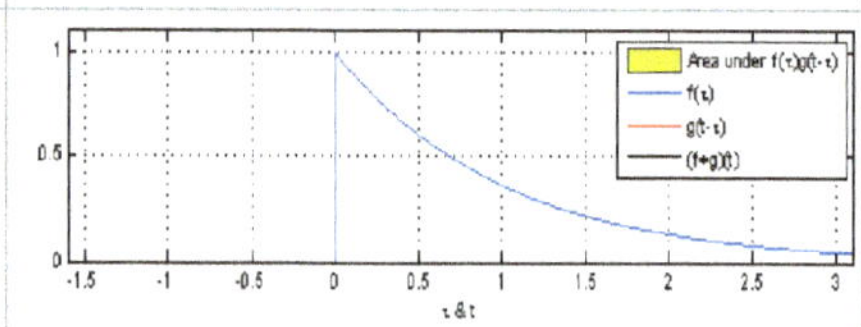

HISTORICAL DEVELOPMENTS

One of the earliest uses of the convolution integral appeared in D'Alembert's derivation of Taylor's theorem in Recherches sur différents points importants du système du monde, published in 1754.

Also, an expression of the type:

$$\int f(u) \cdot g(x-u)\, du$$

is used by Sylvestre François Lacroix on page 505 of his book entitled Treatise on differences and series, which is the last of 3 volumes of the encyclopedic series: Traité du calcul différentiel et du calcul intégral, Chez Courcier, Paris, 1797-1800. Soon thereafter, convolution operations appear in the works of Pierre Simon Laplace, Jean Baptiste Joseph Fourier, Siméon Denis Poisson, and others. The term itself did not come into wide use until the 1950s or 60s. Prior to that it was sometimes known as faltung (which means folding in German), composition product, superposition integral, and Carson's integral. Yet it appears as early as 1903, though the definition is rather unfamiliar in older uses.

The operation:

$$\int_0^t \varphi(s)\psi(t-s)\,ds, \qquad 0 \le t < \infty,$$

is a particular case of composition products considered by the Italian mathematician Vito Volterra in 1913.

CIRCULAR CONVOLUTION

When a function g_T is periodic, with period T, then for functions, f, such that $f*g_T$ exists, the convolution is also periodic and identical to:

$$(f * g_T)(t) \equiv \int_{t_0}^{t_0+T} \left[\sum_{k=-\infty}^{\infty} f(\tau + kT)\right] g_T(t-\tau)\,d\tau,$$

where t_0 is an arbitrary choice. The summation is called a periodic summation of the function f.

When g_T is a periodic summation of another function, g, then $f*g_T$ is known as a circular or cyclic convolution of f and g. And if the periodic summation above is replaced by f_T, the operation is called a periodic convolution of f_T and g_T.

Discrete convolution

For complex-valued functions f, g defined on the set **Z** of integers, the **discrete convolution** of f and g is given by:

$$(f*g)[n] \stackrel{\text{def}}{=} \sum_{m=-\infty}^{\infty} f[m]\,g[n-m]$$

(commutativity)

$$= \sum_{m=-\infty}^{\infty} f[n-m]\,g[m].$$

The convolution of two finite sequences is defined by extending the sequences to finitely supported functions on the set of integers. When the sequences are the coefficients of two polynomials, then the coefficients of the ordinary product of the two polynomials are the convolution of the original two sequences. This is known as the Cauchy product of the coefficients of the sequences.

Thus when g has finite support in the set $\{-M, -M+1, \dots, M-1, M\}$ (representing, for instance, a finite impulse response), a finite summation may be used:

$$(f*g)[n] = \sum_{m=-M}^{M} f[n-m]g[m].$$

CIRCULAR DISCRETE CONVOLUTION

When a function g_N is periodic, with period N, then for functions, f, such that $f*g_N$ exists, the convolution is also periodic and identical to:

$$(f * g_N)[n] \equiv \sum_{m=0}^{N-1} \left(\sum_{k=-\infty}^{\infty} f[m + kN] \right) g_N[n - m].$$

The summation on k is called a periodic summation of the function f.

If g_N is a periodic summation of another function, g, then $f*g_N$ is known as a circular convolution of f and g.

When the non-zero durations of both f and g are limited to the interval [0, N − 1], $f*g_N$ reduces to these common forms:

$$\begin{aligned}(f * g_N)[n] &= \sum_{m=0}^{N-1} f[m]\, g_N[n-m] \\ &= \sum_{m=0}^{n} f[m]\, g[n-m] + \sum_{m=n+1}^{N-1} f[m]\, g[N+n-m] \\ &= \sum_{m=0}^{N-1} f[m]\, g[(n-m)_{\bmod N}] \equiv (f *_N g)[n]\end{aligned}$$

The notation ($f *_N g$) for cyclic convolution denotes convolution over the cyclic group of integers modulo N.

Circular convolution arises most often in the context of fast convolution with an FFT algorithm.

Fast Convolution Algorithms

In many situations, discrete convolutions can be converted to circular convolutions so that fast transforms with a convolution property can be used to implement the computation. For example, convolution of digit sequences is the kernel operation in multiplication of multi-digit numbers, which can therefore be efficiently implemented with transform techniques (Knuth 1997, §4.3.3.C; von zur Gathen & Gerhard 2003, §8.2).

Eq.1 requires N arithmetic operations per output value and N^2 operations for N outputs. That can be significantly reduced with any of several fast algorithms. Digital signal processing and other applications typically use fast convolution algorithms to reduce the cost of the convolution to O(N log N) complexity.

The most common fast convolution algorithms use fast Fourier transform (FFT) algorithms via the circular convolution theorem. Specifically, the circular convolution of two finite-length sequences is found by taking an FFT of each sequence, multiplying pointwise, and then performing an inverse FFT. Convolutions of the type defined above are then efficiently implemented using that technique in conjunction with zero-extension and/or discarding portions of the output. Other fast convolution algorithms, such as the Schönhage–Strassen algorithm or the Mersenne transform, use fast Fourier transforms in other rings.

If one sequence is much longer than the other, zero-extension of the shorter sequence and fast circular convolution is not the most computationally efficient method available. Instead, decomposing the longer sequence into blocks and convolving each block allows for faster algorithms such as the Overlap–save method and Overlap–add method. A hybrid convolution method that combines block and FIR algorithms allows for a zero input-output latency that is useful for real-time convolution computations.

Domain of Definition

The convolution of two complex-valued functions on $\mathbf{R}^d$ is itself a complex-valued function on $\mathbf{R}^d$, defined by:

$$(f*g)(x) = \int_{\mathbf{R}^d} f(y)g(x-y)\,dy = \int_{\mathbf{R}^d} f(x-y)g(y)\,dy,$$

is well-defined only if f and g decay sufficiently rapidly at infinity in order for the integral to exist. Conditions for the existence of the convolution may be tricky, since a blow-up in g at infinity can be easily offset by sufficiently rapid decay in f. The question of existence thus may involve different conditions on f and g:

Compactly Supported Functions

If f and g are compactly supported continuous functions, then their convolution exists, and is also compactly supported and continuous (Hörmander 1983, Chapter 1). More generally, if either function (say f) is compactly supported and the other is locally integrable, then the convolution f∗g is well-defined and continuous.

Convolution of f and g is also well defined when both functions are locally square integrable on **R** and supported on an interval of the form [a, +∞) (or both supported on [-∞, a]).

Integrable Functions

The convolution of f and g exists if f and g are both Lebesgue integrable functions in $L^1(\mathbf{R}^d)$, and in this case f∗g is also integrable (Stein & Weiss 1971, Theorem 1.3). This is a consequence of Tonelli's theorem. This is also true for functions in ℓ^1, under the discrete convolution, or more generally for the convolution on any group.

Likewise, if $f \in L^1(\mathbf{R}^d)$ and $g \in L^p(\mathbf{R}^d)$ where $1 \le p \le \infty$, then $f*g \in L^p(\mathbf{R}^d)$ and

$$\|f*g\|_p \le \|f\|_1 \|g\|_p.$$

In the particular case p = 1, this shows that L^1 is a Banach algebra under the convolution (and equality of the two sides holds if f and g are non-negative almost everywhere).

More generally, Young's inequality implies that the convolution is a continuous bilinear map between suitable L^p spaces. Specifically, if $1 \le p,q,r \le \infty$ satisfy

$$\frac{1}{p}+\frac{1}{q}=\frac{1}{r}+1,$$

then

$$\|f*g\|_r \le \|f\|_p \|g\|_q, \quad f \in \mathcal{L}^p,\ g \in \mathcal{L}^q,$$

so that the convolution is a continuous bilinear mapping from $L^p \times L^q$ to L^r. The Young inequality for convolution is also true in other contexts (circle group, convolution on **Z**). The preceding inequality is not sharp on the real line: when $1 < p, q, r < \infty$, there exists a constant $B_{p,q} < 1$ such that:

$$\|f*g\|_r \le B_{p,q} \|f\|_p \|g\|_q, \quad f \in \mathcal{L}^p,\ g \in \mathcal{L}^q.$$

The optimal value of $B_{p,q}$ was discovered in 1975.

A stronger estimate is true provided $1 < p, q, r < \infty$:

$$\|f*g\|_r \le C_{p,q} \|f\|_p \|g\|_{q,w}$$

where $\|g\|_{q,w}$ is the weak L^q norm. Convolution also defines a bilinear continuous map $L^{p,w} \times L^{q,w} \to L^{r,w}$ for $1 < p, q, r < \infty$, owing to the weak Young inequality:

$$\|f*g\|_{r,w} \le C_{p,q} \|f\|_{p,w} \|g\|_{r,w}.$$

Functions of Rapid Decay

In addition to compactly supported functions and integrable functions, functions that have sufficiently rapid decay at infinity can also be convolved. An important feature of the convolution is that if f and g both decay rapidly, then f∗g also decays rapidly. In particular, if f and g are rapidly decreasing functions, then so is the convolution f∗g. Combined with the fact that convolution commutes with differentiation (see **Properties**), it follows that the class of Schwartz functions is closed under convolution (Stein & Weiss 1971, Theorem 3.3).

Distributions

Under some circumstances, it is possible to define the convolution of a function with a distribution, or of two distributions. If f is a compactly supported function and g is a distribution, then f∗g is a smooth function defined by a distributional formula analogous to

$$\int_{\mathbf{R}^d} f(y)g(x-y)\,dy.$$

More generally, it is possible to extend the definition of the convolution in a unique way so that the associative law

$$f * (g * \varphi) = (f * g) * \varphi$$

remains valid in the case where f is a distribution, and g a compactly supported distribution (Hörmander 1983, §4.2).

Measures

The convolution of any two Borel measures μ and ν of bounded variation is the measure λ defined by (Rudin 1962)

$$\int_{\mathbf{R}^d} f(x)d\lambda(x) = \int_{\mathbf{R}^d}\int_{\mathbf{R}^d} f(x+y)\,d\mu(x)\,d\nu(y).$$

This agrees with the convolution defined above when μ and ν are regarded as distributions, as well as the convolution of L^1 functions when μ and ν are absolutely continuous with respect to the Lebesgue measure.

The convolution of measures also satisfies the following version of Young's inequality

$$\|\mu * \nu\| \le \|\mu\|\|\nu\|$$

where the norm is the total variation of a measure. Because the space of measures of bounded variation is a Banach space, convolution of measures can be treated with standard methods of functional analysis that may not apply for the convolution of distributions.

PROPERTIES

Algebraic Properties

The convolution defines a product on the linear space of integrable functions. This product satisfies the following algebraic properties, which formally mean that the space of integrable functions with the product given by convolution is a commutative algebra without identity (Strichartz 1994, §3.3). Other linear spaces of functions, such as the space of continuous functions of compact support, are closed under the convolution, and so also form commutative algebras.

Commutativity

$$f * (g * h) = (f * g) * h$$

Associativity

$$f * (g * h) = (f * g) * h$$

Distributivity

$$f * (g + h) = (f * g) + (f * h)$$

Associativity with scalar multiplication

$$a(f * g) = (af) * g$$

for any real (or complex) number a.

Multiplicative identity

No algebra of functions possesses an identity for the convolution. The lack of identity is typically not a major inconvenience, since most collections of functions on which the convolution is performed can be convolved with a delta distribution or, at the very least (as is the case of L^1) admit approximations to the identity. The linear space of compactly supported distributions does, however, admit an identity under the convolution. Specifically,

$$f * \delta = f$$

where δ is the delta distribution.

Inverse Element

Some distributions have an inverse element for the convolution, $S^{(-1)}$, which is defined by $S^{(-1)} * S = \delta.$

The set of invertible distributions forms an abelian group under the convolution.

Complex conjugation

$$\overline{f * g} = \overline{f} * \overline{g}$$

Integration

If f and g are integrable functions, then the integral of their convolution on the whole space is simply obtained as the product of their integrals:

$$\int_{\mathbf{R}^d} (f * g)(x)\, dx = \left(\int_{\mathbf{R}^d} f(x)\, dx\right) \left(\int_{\mathbf{R}^d} g(x)\, dx\right).$$

This follows from Fubini's theorem. The same result holds if f and g are only assumed to be nonnegative measurable functions, by Tonelli's theorem.

Differentiation

In the one-variable case,

$$\frac{d}{dx}(f*g)=\frac{df}{dx}*g=f*\frac{dg}{dx}$$

where d/dx is the derivative. More generally, in the case of functions of several variables, an analogous formula holds with the partial derivative:

$$\frac{\partial}{\partial x_i}(f*g)=\frac{\partial f}{\partial x_i}*g=f*\frac{\partial g}{\partial x_i}.$$

A particular consequence of this is that the convolution can be viewed as a "smoothing" operation: the convolution of f and g is differentiable as many times as f and g are in total.

These identities hold under the precise condition that f and g are absolutely integrable and at least one of them has an absolutely integrable (L^1) weak derivative, as a consequence of Young's inequality. For instance, when f is continuously differentiable with compact support, and g is an arbitrary locally integrable function,

$$\frac{d}{dx}(f*g)=\frac{df}{dx}*g.$$

These identities also hold much more broadly in the sense of tempered distributions if one of f or g is a compactly supported distribution or a Schwartz function and the other is a tempered distribution. On the other hand, two positive integrable and infinitely differentiable functions may have a nowhere continuous convolution.

In the discrete case, the difference operator D f(n) = f(n + 1) − f(n) satisfies an analogous relationship:

$$D(f*g)=(Df)*g=f*(Dg).$$

Convolution the Orem

The convolution theorem states that

$$\mathcal{F}\{f*g\}=k\cdot\mathcal{F}\{f\}\cdot\mathcal{F}\{g\}$$

where $\mathcal{F}\{f\}$ denotes the Fourier transform of f, and k is a constant that depends on the specific normalization of the Fourier transform. Versions of this theorem also hold for the Laplace transform, two-sided Laplace transform, Z-transform and Mellin transform.

Translation Invariance

The convolution commutes with translations, meaning that

$$\tau_x(f*g)=(\tau_x f)*g=f*(\tau_x g)$$

where $\tau_x f$ is the translation of the function f by x defined by

$(\tau_x f)(y) = f(y - x).$

If f is a Schwartz function, then $\tau_x f$ is the convolution with a translated Dirac delta function $\tau_x f = f * \tau_x \delta$. So translation invariance of the convolution of Schwartz functions is a consequence of the associativity of convolution.

Furthermore, under certain conditions, convolution is the most general translation invariant operation. Informally speaking, the following holds

- Suppose that S is a linear operator acting on functions which commutes with translations: $S(\tau_x f) = \tau_x(Sf)$ for all x. Then S is given as convolution with a function (or distribution) g_S; that is $Sf = g_S * f$.

Thus any translation invariant operation can be represented as a convolution. Convolutions play an important role in the study of time-invariant systems, and especially LTI system theory. The representing function g_S is the impulse response of the transformation S.

A more precise version of the theorem quoted above requires specifying the class of functions on which the convolution is defined, and also requires assuming in addition that S must be a continuous linear operator with respect to the appropriate topology. It is known, for instance, that every continuous translation invariant continuous linear operator on L^1 is the convolution with a finite Borel measure. More generally, every continuous translation invariant continuous linear operator on L^p for $1 \leq p < \infty$ is the convolution with a tempered distribution whose Fourier transform is bounded. To wit, they are all given by bounded Fourier multipliers.

Convolutions on groups

If G is a suitable group endowed with a measure λ, and if f and g are real or complex valued integrable functions on G, then we can define their convolution by

$$(f * g)(x) = \int_G f(y)g(y^{-1}x)\, d\lambda(y).$$

It is not commutative in general. In typical cases of interest G is a locally compact Hausdorff topological group and λ is a (left-) Haar measure. In that case, unless G is unimodular, the convolution defined in this way is not the same as $\int f(xy^{-1})g(y)\, d\lambda(y)$. The preference of one over the other is made so that convolution with a fixed function g commutes with left translation in the group:

$$L_h(f * g) = (L_h f) * g.$$

Furthermore, the convention is also required for consistency with the definition of the convolution of measures given below. However, with a right instead of a left Haar measure, the latter integral is preferred over the former.

On locally compact abelian groups, a version of the convolution theorem holds: the Fourier transform of a convolution is the pointwise product of the Fourier transforms. The circle group **T** with the Lebesgue measure is an immediate example. For a fixed g in $L^1(\mathbf{T})$, we have the following familiar operator acting on the Hilbert space $L^2(\mathbf{T})$:

$$Tf(x) = \frac{1}{2\pi}\int_{\mathbf{T}} f(y)g(x-y)\,dy.$$

The operator T is compact. A direct calculation shows that its adjoint T* is convolution with

$$\bar{g}(-y).$$

By the commutativity property cited above, T is normal: T*T = TT*. Also, T commutes with the translation operators. Consider the family S of operators consisting of all such convolutions and the translation operators. Then S is a commuting family of normal operators. According to spectral theory, there exists an orthonormal basis $\{h_k\}$ that simultaneously diagonalizes S. This characterizes convolutions on the circle. Specifically, we have

$$h_k(x) = e^{ikx}, \quad k \in \mathbb{Z},$$

which are precisely the characters of **T**. Each convolution is a compact multiplication operator in this basis. This can be viewed as a version of the convolution theorem discussed above.

A discrete example is a finite cyclic group of order n. Convolution operators are here represented by circulant matrices, and can be diagonalized by the discrete Fourier transform.

A similar result holds for compact groups (not necessarily abelian): the matrix coefficients of finite-dimensional unitary representations form an orthonormal basis in L^2 by the Peter–Weyl theorem, and an analog of the convolution theorem continues to hold, along with many other aspects of harmonic analysis that depend on the Fourier transform.

Convolution of measures

Let G be a topological group. If μ and ν are finite Borel measures on G, then their convolution μ∗ν is defined by

$$(\mu * \nu)(E) = \iint 1_E(xy)\,d\mu(x)\,d\nu(y)$$

for each measurable subset E of G. The convolution is also a finite measure, whose total variation satisfies

$$\|\mu * \nu\| \le \|\mu\|\|\nu\|.$$

In the case when G is locally compact with (left-)Haar measure λ, and μ and ν are absolutely continuous with respect to a λ, so that each has a density function, then the convolution μ∗ν is also absolutely continuous, and its density function is just the convolution of the two separate density functions.

If μ and ν are probability measures on the topological group (**R**,+), then the convolution μ∗ν is the probability distribution of the sum X + Y of two independent random variables X and Y whose respective distributions are μ and ν.

BIALGEBRAS

Let (X, Δ, ∇, ε, η) be a bialgebra with comultiplication Δ, multiplication ∇, unit η, and counit ε. The convolution is a product defined on the endomorphism algebra End(X) as follows. Let φ, ψ ∈ End(X), that is, φ,ψ : X → X are functions that respect all algebraic structure of X, then the convolution φ∗ψ is defined as the composition

$$X \xrightarrow{\Delta} X \otimes X \xrightarrow{\phi \otimes \psi} X \otimes X \xrightarrow{\nabla} X.$$

The convolution appears notably in the definition of Hopf algebras (Kassel 1995, §III.3). A bialgebra is a Hopf algebra if and only if it has an antipode: an endomorphism S such that

$$S * \mathrm{id}_X = \mathrm{id}_X * S = \eta \circ \varepsilon.$$

APPLICATIONS

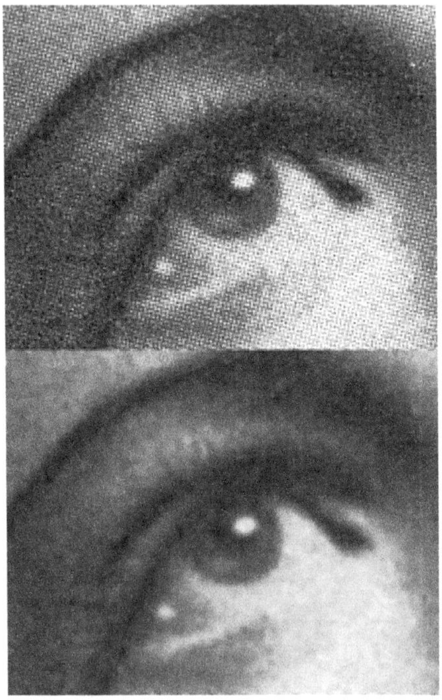

Gaussian blur can be used in order to obtain a smooth grayscale digital image of a halftone print

Convolution and related operations are found in many applications in science, engineering and mathematics.

- In image processing

In digital image processing convolutional filtering plays an important role in many important algorithms in edge detection and related processes.

In optics, an out-of-focus photograph is a convolution of the sharp image with a lens function. The photographic term for this is bokeh.

In image processing applications such as adding blurring.

- In digital data processing

In analytical chemistry, Savitzky–Golay smoothing filters are used for the analysis of spectroscopic data. They can improve signal-to-noise ratio with minimal distortion of the spectra.

In statistics, a weighted moving average is a convolution.

- In acoustics, reverberation is the convolution of the original sound with echoes from objects surrounding the sound source.

In digital signal processing, convolution is used to map the impulse response of a real room on a digital audio signal.

In electronic music convolution is the imposition of a spectral or rhythmic structure on a sound. Often this envelope or structure is taken from another sound. The convolution of two signals is the filtering of one through the other.

- In electrical engineering, the convolution of one function (the input signal) with a second function (the impulse response) gives the output of a linear time-invariant system (LTI). At any given moment, the output is an accumulated effect of all the prior values of the input function, with the most recent values typically having the most influence (expressed as a multiplicative factor). The impulse response function provides that factor as a function of the elapsed time since each input value occurred.
- In physics, wherever there is a linear system with a "superposition principle, a convolution operation makes an appearance. For instance, in spectroscopy line broadening due to the Doppler effect on its own gives a Gaussian spectral line shape and collision broadening alone gives a Lorentzian line shape. When both effects are operative, the line shape is a convolution of Gaussian and Lorentzian, a Voigt function.

In Time-resolved fluorescence spectroscopy, the excitation signal can be treated as a chain of delta pulses, and the measured fluorescence is a sum of exponential decays from each delta pulse.

In computational fluid dynamics, the large eddy simulation (LES) turbulence model uses the convolution operation to lower the range of length scales necessary in computation thereby reducing computational cost.

- In probability theory, the probability distribution of the sum of two independent random variables is the convolution of their individual distributions.

In kernel density estimation, a distribution is estimated from sample points by convolution with a kernel, such as an isotropic Gaussian. (Diggle 1995).

- In radiotherapy treatment planning systems, most part of all modern codes of calculation applies a convolution-superposition algorithm.
- Convolutional neural networks apply multiple cascaded convolution kernels with applications in machine vision and artificial intelligence

REFERENCES

1. Bracewell, R. (1986), The Fourier Transform and Its Applications (2nd ed.), McGraw–Hill, ISBN 0-07-116043-4.
2. Damelin, S.; Miller, W. (2011), The Mathematics of Signal Processing, Cambridge University Press, ISBN 978-1107601048
3. Diggle, P. J., "A kernel method for smoothing point process data", Journal of the Royal Statistical Society, Series C, **34**: 138–147, doi:10.2307/2347366
4. Dominguez-Torres, Alejandro (Nov 2, 2010). "Origin and history of convolution". 41 pgs. http://www.slideshare.net/Alexdfar/origin-adn-history-of-convolution. Cranfield, Bedford MK43 OAL, UK. Retrieved Mar 13, 2013.
5. Hewitt, Edwin; Ross, Kenneth A. (1979), Abstract harmonic analysis. Vol. I, Grundlehren der Mathematischen Wissenschaften [Fundamental Principles of Mathematical Sciences], **115** (2nd ed.), Berlin, New York: Springer-Verlag, ISBN 978-3-540-09434-0, MR 551496.
6. Hewitt, Edwin; Ross, Kenneth A. (1970), Abstract harmonic analysis. Vol. II: Structure and analysis for compact groups. Analysis on locally compact Abelian groups, Die Grundlehren der mathematischen Wissenschaften, Band 152, Berlin, New York: Springer-Verlag, MR 0262773.
7. Hörmander, L. (1983), The analysis of linear partial differential operators I, Grundl. Math. Wissenschaft., **256**, Springer, ISBN 3-540-12104-8, MR 0717035.
8. Kassel, Christian (1995), Quantum groups, Graduate Texts in Mathematics, **155**, Berlin, New York: Springer-Verlag, ISBN 978-0-387-94370-1, MR 1321145.
9. Knuth, Donald (1997), Seminumerical Algorithms (3rd. ed.), Reading, Massachusetts: Addison–Wesley, ISBN 0-201-89684-2.
10. Reed, Michael; Simon, Barry (1975), Methods of modern mathematical physics. II. Fourier analysis, self-adjointness, New York-London: Academic Press Harcourt Brace Jovanovich, Publishers, pp. xv+361, ISBN 0-12-585002-6, MR 0493420
11. Rudin, Walter (1962), Fourier analysis on groups, Interscience Tracts in Pure and Applied Mathematics, No. 12, Interscience Publishers (a division of John Wiley and Sons), New York–London, ISBN 0-471-52364-X, MR 0152834.
12. Sobolev, V.I. (2001), Convolution of functions, in Hazewinkel, Michiel, Encyclopedia of Mathematics, Springer, ISBN 978-1-55608-010-4.
13. Stein, Elias; Weiss, Guido (1971), Introduction to Fourier Analysis on Euclidean Spaces, Princeton University Press, ISBN 0-691-08078-X.
14. Strichartz, R. (1994), A Guide to Distribution Theory and Fourier Transforms, CRC Press, ISBN 0-8493-8273-4.
15. Titchmarsh, E (1948), Introduction to the theory of Fourier integrals (2nd ed.), New York, N.Y.: Chelsea Pub. Co. (published 1986), ISBN 978-0-8284-0324-5.
16. Uludag, A. M. (1998), On possible deterioration of smoothness under the operation of convolution, J. Math. Anal. Appl. 227 no. 2, 335–358

17. Treves, François (1967), Topological Vector Spaces, Distributions and Kernels, Academic Press, ISBN 0-486-45352-9.
18. von zur Gathen, J.; Gerhard, J. (2003), Modern Computer Algebra, Cambridge University Press, ISBN 0-521-82646-2.

Index